MULTIPLE CLASSIFICATION ANALYSIS

A REPORT ON A COMPUTER PROGRAM FOR MULTIPLE REGRESSION USING CATEGORICAL PREDICTORS

Frank M. Andrews
James N. Morgan
John A. Sonquist
Laura Klem

SECOND EDITION

ISR Code No. 2572-73

Library of Congress Catalog Card No. 73-620206
ISBN 0-87944-055-4 paperbound
ISBN 0-87944-148-8 clothbound

Published by the Institute for Social Research
The University of Michigan, Ann Arbor, Michigan 48106

First Published 1967
Second Edition, Revised 1973
© 1973 The University of Michigan, All Rights Reserved
Manufactured in the United States of America

Table of Contents

LIST OF FIGURES

Preface to the Second Edition

The First Edition of this monograph appeared in 1967 and went through five printings. During the years following 1967, various changes were incorporated in the Multiple Classification Analysis program as research staff at the Institute for Social Research found ways to improve it and as it was adapted to run on computing hardware which was itself evolving. In the process, successive printings of the original monograph became increasingly obsolete.

This Second Edition represents a significant up-dating of the original monograph, involving substantial changes in the text and modest expansion. Laura Klem, who has played the major role in revising the original work, appears as co-author of this Second Edition.

If we can believe what we hear from our mail, phone calls, and visitors, the Multiple Classification Analysis program has proved useful to a wide variety of data analysts in the social sciences and other fields. We are grateful for the many suggestions we have received for ways to improve the program itself, or the monograph which describes it. Our own learning and thinking has been stimulated by some of the questions we have been asked to answer. We have tried to incorporate many of these suggestions and learnings in this Second Edition and hope its readers will find it a useful and understandable document.

Acknowledgments:
The Development of the Program

The Survey Research Center's Multiple Classification Analysis program in its present form is a product of many groups and individuals. It is a pleasure to cite some of the important contributors.

The present program derives from an early version written under instructions of Vernon Lippitt for the IBM 650 at the General Electric Company (Lippitt 1959). A copy of this program was received in 1959 by the Economic Behavior Program of the Survey Research Center and made operational by Edwin Dean. Using formulas provided by Irene Hess and R.K. Pillai, an improved Fortran version was programmed for the University of Michigan's IBM 704.

Several improvements to the Multiple Classification Analysis program have been made since then. The program discussed in the first edition of this monograph was written in Fortran II for the IBM 7090 and was designed to run under the University of Michigan Executive System. The version discussed here is written in Fortran IV and 360 Assembler for the IBM 360 series of computers and is designed to run under OS/360. This version of the program maintains the original algorithm but provides substantial improvements in input and output capabilities, including an option for computation of predicted values and residuals. Sylvia Barge, David Kapell, Robert C. Messenger, Judith Rattenbury and Peter Solenberger have been responsible for the conversion and improvements.

The task of discovering what the program would do under various circumstances and documenting these discoveries has itself required substantial effort. This monograph draws heavily on a series of earlier documents to which Donald C. Pelz and Barbara Baldwin, in addition to the present authors, were major contributors.

These various operations have, of course, required both money and computer time. We are grateful to Dr. R.C.F. Bartels, Director of The University of Michigan Computing Center, for providing access to the series of computers that have been under his charge. We are also grateful to the National Science Foundation, whose funds occasionally supported the development of this program when we had no choice but to solve methodological problems before we could proceed with substantive investigations. A Faculty Research Grant from the Horace H. Rackham School of Graduate Studies at The University of Michigan supported the preparation of the first edition of this monograph. The Survey Research Center Computer Support Group supported the preparation of this revised edition.

Finally, we are grateful to Mary Hope, Mildred Dennis and Juanita Wheeler for secretarial help in preparing the various versions of this document.

Introduction

Multiple Classification Analysis (MCA) is a complex technique for multivariate analysis which has proved extremely useful at the Survey Research Center. Colleagues and students, both at the University of Michigan and elsewhere, have sought information about the program with increasing frequency. This monograph is intended to answer their many questions. It has two major parts.

Part I is addressed to the potential user who wants to know whether the technique is appropriate for a particular problem, what the program can do, and how one may interpret its results. Chapter 1 describes the problems for which MCA is particularly appropriate, and the answers it can give. Chapter 2 describes the major features of the MCA program. Chapter 3 discusses the limitations of the program, problems the user may encounter, and how such problems may be solved. Readers who are knowledgable about multivariate techniques may wish only to skim Chapters 1 and 3; Chapter 2, on the other hand, contains details specific to the MCA program.

Part II is more statistical. Chapter 4 presents the additive model on which the algorithm is based and the formulas which generate the program output. Chapters 5 and 6 respectively, relate Multiple Classification Analysis to more traditional analysis of variance and multiple regression techniques.

A set of Appendices describes the present program. They give exact set-up instructions, a guide to the printout, a flow chart and sample output. An Appendix on obtaining the program and adapting it to another computer is included.

A Bibliography, which includes examples of substantive studies which have made use of the MCA program, completes the monograph.

Part One

A Non-Technical Description of Multiple Classification Analysis

1

Uses of Multiple Classification Analysis

1.0 Summary

Multiple Classification Analysis (MCA) is a technique for examining the interrelationships between several predictor variables and a dependent variable within the context of an additive model. Unlike simpler forms of other multivariate methods, the technique can handle predictors with no better than nominal measurement, and interrelationships of any form among predictors or between a predictor and the dependent variable. The dependent variable, however, should be an intervally scaled (or a numerical) variable without extreme skewness, or a dichotomous variable with two frequencies which are not extremely unequal.

The statistics printed by the program show how each predictor relates to the dependent variable, both before and after adjusting for the effects of other predictors, and how all the predictors considered together relate to the dependent variable.

A discussion is presented below of the problems correlated predictors pose for multivariate methods; an analysis problem is presented which will be used as an example throughout the monograph.

1.1 Some Goals of Multivariate Analysis

Many of the most interesting analysis problems involve the simultaneous consideration of several predictor variables (i.e., "independent" variables) and their relationships to a dependent variable. Sometimes one wants to know *how well* all the variables together explain variation in the dependent variable. Other times it is necessary to look at each predictor separately to see how it relates to the dependent variable, either considering or neglecting the effects of other predictors. A criterion used generally is its contribution to reduction in unexplained variance or "error." Another is the extent to which its class means differ from the grand mean.

A related, but different, concern is the matter of predicted relations. Instead of asking *how well* one can predict, one sometimes asks *what level*

1

(i.e., what particular value or score) would one predict for a person or other unit having a certain combination of characteristics. This is the classic problem to which multiple regression has frequently been applied.

Finally, one sometimes wants to know whether one's ability to predict is *significantly better than chance*. Tests of significance (F,*t*) are the usual criteria.

The MCA program implements a multivariate technique that is relevant for all of the above problems and that may be applied to many kinds of data for which the simpler forms of the traditional techniques would be inappropriate. Weak measurement (including nominal scales) on the predictor variables, correlated predictors, and non-linear relationships are conditions which the MCA program is designed to handle. In essence it is multiple regression using dummy variables (Suits 1957; Melichar 1965). Its chief advantage over conventional dummy variable regression is a more convenient input arrangement and understandable output that focuses on *sets* of predictors such as occupation groups, and on the extent and direction of the adjustments made for intercorrelations among the sets of predictors.

1.2 A Multivariate Analysis Problem

Consider the problem of predicting the performance of scientists in a laboratory. Of course, numerous factors may affect it, but here we shall consider only three: amount of formal training, amount of professional experience, and amount of seniority in the laboratory. These were three predictors in an analysis conducted by Pelz and Andrews (1966). Output of reports (expressed on a percentile scale) over the past five years was the dependent variable.

Scientists were divided into three categories on the first predictor, education, according to the highest academic degree they held, BS, MS, or PhD. On the second predictor, they were classified into four categories depending on the number of years which had elapsed since they received their highest academic degree: 0-2 years, 3-5, 6-10, and 11 or more. On the third predictor, classification was into one of five categories based on the number of years they had worked for their present employer, 0-1 years, 2, 3-4, 5-9, and 10 or more.

It was no surprise to discover that these predictors were themselves interrelated. In general, the longer the experience, the greater the seniority. Furthermore, due to a change in hiring policy, formal education was also related to the other two predictors: scientists with short experience and seniority were more likely to have PhD's than were other scientists.

Finally, it should be noted that while report production was expected to relate positively to seniority, experience, and highest degree, the relationships were not expected to be linear. Increases in seniority and experience were expected to be accompanied by the sharpest increases in performance during the early years. It was expected that scientists with BS or MS degrees might be performing at about the same levels, but that PhD's would be substantially above either of these groups.

Here, then, were data of which several questions might be asked: How strong was the relationship between report production and amount of formal training, professional experience, seniority? How strong would these relationships remain if the effects of the other predictors were taken into account? How well could report production be explained by all three predictors together? Were these relationships strong enough to be considered statistically significant? What level of reports might be expected for a person with given amounts of training, experience, and seniority? These are typical of the questions one asks in multivariate analysis.

An analysis technique that goes one step beyond the determination of expected levels for people with given characteristics is the examination of "residuals." By taking a scientist's score on the dependent variable and subtracting his expected score (i.e., the output which would be expected for people with his characteristics), one obtains a residual score indicating how much above or below normal that person was. When this operation is performed for each case in the analysis, a new dependent variable is obtained, from which all "additive" effects attributable to the predictors have been removed or "partialed out." It may then be of considerable interest to see what accounts for the remaining variability, particularly if the distribution is not normal. (In the above example, Pelz and Andrews determined whether certain social or psychological factors affected the likelihood of a scientist's performance being above or below the expected levels for scientists with his education, seniority, and experience.) There is, of course, a downward bias in the estimated effects of the variables used at the second stage, the size of the bias being proportional to the correlation between the second stage predictors and the first stage ones (Morgan 1958; Goldberger and Jochem 1961). Similarly, there is a downward bias in the estimate of how much variance is explained by the predictors used at the second stage.

1.3 Special Advantages of the MCA Technique

The MCA technique overcomes some of the problems of attempting to apply either of the two more usual multivariate procedures to survey data. If analysis of variance is to be used, the problem of correlated predictors must be considered. If multiple regression or discriminant function analysis is to be used, one is faced with the problem of predictors which are not numerical variables but categories, often with scales as weak as the nominal level. [1]

---

[1] We will classify scales into nominal, ordinal, and interval levels of measurement. This distinction was first proposed by S. S. Stevens (1946) and is discussed in many textbooks. A nominal

Let us consider first the problem of correlated predictors. Traditional analysis of variance techniques generally require that the predictors be independent. This is often expressed in terms of "equal or proportional numbers of cases in the cells." A scientist generally designs an experiment to achieve this. He acquires "experimental control" by causing his predictors to be unrelated. When observations are not subject to human control, however, it usually happens that the predictors prove to be correlated with one another. In this case, the analysis of the resulting data is more complicated.

Because dealing with this complication is one key feature of the MCA program, we consider our earlier example further. It happened that 8 percent of the variation in scientists' report production could be attributed to differences in their professional experience. Similarly, differences in seniority accounted for 15 percent of the variation in their performance. Did this mean that one could "explain" 23 percent (8 percent + 15 percent) of the performance variation using these two predictors? Probably not, for part of the variation explained by one predictor could *also* be explained by the other predictor. We noted that the predictors were positively related to one another. In a sense they "overlapped," hence the two considered together would explain *less* than the sum of the two, each considered separately. Figure 1 presents this in schematic form. The total area covered by the two circles is less than the sum of their individual areas because they overlap.

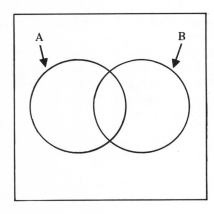

FIGURE 1
CORRELATED PREDICTORS
The total area within the box represents the variation of the dependent variable. The area enclosed within circle A shows the portion of this variation which can be explained by Predictor A. The area within circle B shows the variation attributable to predictor B. The total enclosed area is less than the sum of the areas enclosed by the two circles separately since the circles overlap—i.e., the predictors are correlated.

---

scale is one which simply categorizes objects (e.g., apples, oranges, pears). An ordinal scale classifies items and assumes the categories are arranged in some meaningful order. An interval scale requires classification, ordering, and equal distances between the categories. Stevens also distinguished a subclass of interval scales which he called ratio scales; we will not make this distinction.

The phenomenon would work in an opposite fashion if the predictors had been negatively related to each other and positively related to the dependent variable. Together they would have explained *more* of the variation in the dependent variable than the sum of the two considered separately. This is the well known "suppressor effect."

Correlated predictors not only make problems for estimating the total variation explained by a set of predictors, they also affect one's estimate of the predicted value of the dependent variable for any particular individual (or unit). It was observed that scientists with 11 or more years of professional experience tended to score 8 percentile points above average on production of reports. Those with 10 or more years of seniority in their lab tended to score 11 percentile points above average. One might be tempted to predict that a scientist with 11 or more years experience *and* 10 or more years seniority would score 19 (8 + 11) percentile points above average.[2]

A moment's reflection, however, will indicate that this is probably wrong. Part of the reason scientists with long experience score above average is that many *also* have long seniority. Thus the apparent effect of long seniority in boosting performance reflects the "true" seniority effect plus some of the effect of long experience which "spills over." In the same way, the apparent experience effect includes both the true experience effect and some of the seniority effect. The amount of "spill over," of course, depends on the degree of relationship between the two predictors. If the relationship is positive, the true effects will in general be less than the apparent effects (assuming both predictors relate to the dependent variable in the same way). Conversely, if the relationship is negative, the true effects will be more than the sum of the apparent effects.

A key feature of the MCA technique is its ability to show the effect of each predictor on the dependent variable both before and after taking into account the effects of all other variables. Of course, simple forms of traditional multivariate methods (analysis of covariance, multiple regression, and discriminant functions) also do this, but they can do it only when the data are of a prescribed form. They usually require that all variables be measured on interval scales, and that relationships be linear (or linearized).

The second key feature of the MCA program lies in its freedom from these restrictions. The predictors are always treated as sets of classes or categories, so it does not matter whether a particular set represents a nominal scale (categories), an ordinal scale (ranking), or an interval scale (classes of a numerical variable).[3] Since the categories of a nominal scale can be placed in

---

[2]This discussion, like the MCA technique itself, assumes an additive model underlying the phenomena. A formal presentation of the model appears in Chapter 4.

[3]In using the MCA program there is no need to generate 1-0 "dummy" variables explicitly. The two-way tables of sets of predictors are available, as are the means of the dependent variable

any order, one cannot speak of the direction or sign of the relation with the dependent variable or even of the form of its relation to other scales.

The usual way of dealing with non-numerical variables in regression is to convert them into a series of dummy variables which take the value of unity if the individual belongs to a particular category of a scale and take the value zero if otherwise. We shall see later that there is a simple relation between the coefficients of such an analysis and the MCA statistics, and that the advantages of using the MCA program are largely those of convenience in input, output, and presentation. The MCA coefficients are all expressed as adjustments to the grand mean, not deviations from the single class which must be excluded from each set when dummy variables are used.

1.4 Data Appropriate for an MCA

From the discussion above, it should be clear that the appropriate data for a MCA consist of one dependent variable and several predictor variables. The dependent variable should be measured on an interval scale or be a dichotomy. It should not be badly skewed. The predictors may be measured on nominal, ordinal, or interval scales. Exact coding requirements for the input variables are given, and distribution requirements are discussed, in the first sections of Chapter 2.

From a statistical standpoint, the technique assumes that the dependent variable is predictable from an additive combination of the predictor variables. Furthermore, it requires a large number of cases. These limitations, plus certain others, are discussed fully in Chapter 3.

1.5 Printout from the Program

This section describes the major statistics printed by the program. The goal is to help the potential user decide whether this program produces what he needs. The printout includes the following items (some of which are optional at the discretion of the user). If sampling weights are used, each item will reflect the weights.

—Bivariate frequency tables for each pair of predictors.
—Number of cases in the analysis.
—Grand mean for the dependent variable.
—Standard deviation of the dependent variable.
—Total, explained, and residual sums of squares
—For each category of each predictor:
　–Number of cases in the category.

---

for each category of each predictor.

It may seem that converting a numerical predictor into a set of classes loses information, but it is easy to demonstrate that as few as five classes of roughly equal numbers of cases generally account for more than ninety percent of the variance the variable could possibly account for, provided the variable has no extreme cases. This is discussed further in section 2.1b.

–Percentage of all cases which fall in the category.

–Mean value of dependent variable for the category, i.e., the raw mean.

–Deviation of raw mean for the category from the grand mean. (This indicates the gross or unadjusted effect of the predictor.)

–Deviation of category mean from grand mean. (This indicates the effect of the predictor after adjusting for effects of other predictors, i.e., after all other predictors have been "held constant.")

–Adjusted mean, i.e., the grand mean plus the deviation of the category mean. (This indicates what the mean would have been if the group had been exactly like the total population with respect to its distribution over all the other predictor classifications.)

–Standard deviation of the dependent variable.

—For each predictor:

–Eta and eta ^{2}. Eta indicates the ability of the predictor, using the categories given, to explain variation in the dependent variable. Eta^2 is the correlation ratio and indicates the proportion of the total sum of squares explainable by the predictor.

–Beta and beta^2. These are directly analogous to the eta statistics, but are based on the adjusted means rather than the raw means. Beta provices a measure of the ability of the predictor to explain variation in the dependent variable after adjusting for the effects of all other predictors. This is *not* in terms of percent of variance explained.[4]

–Sum of squared deviations of the raw means from the grand mean (weighted for number of cases).

–Sum of squared deviations of adjusted means from the grand mean (weighted for number of cases).

—For all predictors considered together:

–A multiple correlation coefficient squared (unadjusted for degrees of freedom). This coefficient indicates the proportion of variance explained by this run of the program.

–Adjustment for degrees of freedom. This is the factor used to correct for captialization on chance in fitting the model in the particular sample being analyzed.

–A multiple correlation coefficient (adjusted for degrees of freedom).

–A multiple correlation coefficient squared (adjusted for degrees of freedom). This coefficient indicates the proportion of variance in the dependent variable explained by all predictors together.

---

[4]The interpretation of beta is discussed in section 6.3. The term beta is used because the measure is analogous to the standardized regression coefficient, i.e., the regression coefficient multiplied by the standard deviation of the predictor and divided by the standard deviation of the dependent variable, so that the result is a measure of the number of standard deviation units the dependent variable moves when the explanatory variable changes by one standard deviation. Ezekiel and Fox call such a measure the "partial beta coefficient" (1959, page 196).

The program does not compute F tests.[5] However the basic statistics from which several different F tests can be easily calculated are part of the output. Formulas for computing two F tests are given in Chapter 4. Also presented in Chapter 4 are procedures for determining the marginal importance of a predictor or group of predictors.

---

[5]The program does not compute F tests when it is used for a Multiple Classification Analysis. There is however a special program option, described in the next chapter, which allows the program to be used for ordinary one-way analysis of variance. When the program is used in this latter way an F value is printed.

2

Major Features of the Program

2.0 Summary

The first sections of this chapter describe in detail the data which are appropriate for MCA analysis and the data management capabilities of the program. Section 2.4 describes the two user options, for a test and a tolerance, which stem from the numerical method used by the program. Section 2.5 describes an option for one-way analysis of variance. The final section describes the option for residuals output.

2.1 Input Data

A MCA analysis requires a dependent variable and two or more predictor variables. The use of a weight variable is a program option.

2.1a. The dependent variable. The dependent variable should be an interval scale. To use the program for analyzing a dependent variable having ordinal properties, one must be willing to assume it approximates an underlying scale.[1] The program may also be used on a two-point nominal scale (e.g. $0 =$ no, $1 =$ yes or $1 =$ no, $5 =$ yes). If the codes 0 and 1 are used, one is in effect using a "proportion" scale in the sense that a mean of .65 for a group would indicate "65 percent yes" and the output statistics are equivalent to a two-group discriminant function analysis.

There is no serious limitation to the range of values the dependent variable may embrace, i.e., it may have as many as seven digits, with any number of decimal places, and may contain either positive or negative values.

Whatever the dependent variable, it should not be extremely skewed.[2] Cases far out on a long "tail" unduly affect the value of the means and variance which are crucial for the calculation of other statistics. Moderate

---

[1] If the dependent variable is measured on a nominal scale, an analysis much like MCA can be done using the MNA program. See Andrews and Messenger (1973).

[2] Lansing and Morgan (1971) discuss ways of handling several different distribution problems. See their Chapter 6, particularly the section on limited dependent variables.

departures from symmetry, however, seem not to pose serious problems. In any case, the program can check for cases larger than a certain prescribed value (which can be set so it is, say, three or four standard deviations from the mean) and can remove these cases before performing the analysis. By specifying a stringent criteria, i.e., a low number of standard deviation, the user can ensure that the data actually used in the analysis are not badly skewed. Alternatively, a transformation of the dependent variable can be used to avoid results dominated by a few cases. For example, replacing the dependent variable by its square root or logarithm may prove useful. If the dependent variable is a dichotomy, the smaller category should contain at least ten percent of the cases.[3]

Cases with missing data on the dependent variable can, and usually should be omitted from the analysis. There is a variety of convenient ways to specify that certain cases should be excluded from an analysis.

2.1b. Predictor variables. There may be from 2 to 200 predictor variables.[4] The number which can be used in a particular analysis depends, however, on how many classes each predictor variable has—see section 3. Experience indicates that results are most easily interpreted if relatively few (i.e., three to ten) predictors are used.

They may be at any level of measurement, including nominal. Each predictor must be represented as a one or two digit code with up to 32 categories. The code values must be in the range 0-31. If a case has a code greater than 31 for any predictor, the case is omitted from the entire run. Although a predictor may have as many as 32 categories, 6 or fewer categories are preferable. Kalton (1966) has shown that one loses very little precision by categorizing a numerical variable into a set of classifications. If, for instance, a predictor has a linear relationship with some dependent variable, accounting for some M percent of its variance, then a categorization into as few as five subclasses of the explanatory variable will account for ninety-five percent of that potential M percent, and ten subclasses will account for ninety-nine percent of the total possible. (For a rectangular distribution, the formula

---

[3]The dichotomous dependent variable causes difficulties when its expected value is close to zero or to one, for two reasons: (a) heteroscedasticity and (b) the possibility that an additive model will produce predictions less than zero or greater than one, which are impossible. The first problems can be solved by weighted regression (rerunning the MCA after weighting each case by the inverse of its variance, which in this case is $1/pq$, where p is the probability predicted from the first MCA). The second requires more complex procedures such as logit or probit analysis, or a better specification of the model to take account of the non-additivities that are causing the trouble.

[4]This section describes requirements for predictors when the MCA program is being used for a Multiple Classification Analysis. As described below, the program can also be used for a one-way analysis of variance; in the latter mode, predictor requirements are somewhat less strict than those given here. See section 2.5.

is $1 - 1/k^2 = $ eta^2 /r^2, where k is the number of subclasses.) And if, in fact, the relationship is not linear, a categorized variable may easily account for more of the total variance than a linear regression using the full numerical detail.

In general it is a good idea to have cases fairly evenly distributed across categories. However if a predictor category has only a few cases it causes little difficulty. Its coefficient may be unstable and also quite large, in either a positive or a negative direction. But although the other coefficients in the set have to deviate the other way to bring the weighted mean of the coefficients to zero, the effect on the other coefficients is small since the weights are so disparate.

Missing data on a predictor should be coded as one or more separate categories. Cases with missing data can be retained in the analysis or the program can be instructed to omit these cases. If the number of predictors is large, it is generally better to leave cases with missing data in the analysis, since otherwise an unnecessarily high percent of cases might be excluded. (Four percent missing data on each of ten predictors *could* result in exclusion of as many as forty percent of the total cases.) A missing data category will probably have only a few cases, and, as noted in the preceding paragraph, a category of a predictor with only a few cases causes little difficulty.

Predictors may be intercorrelated with each other. As noted in section 3.3, however, extremely high correlation between two (or more) predictors may lead to problems. Problems may also result if two categories overlap with respect to the cases in them, e.g., if nearly all the cases which fall in category i of predictor 1 also fall in category j of predictor 2. Problems are sure to occur if one includes predictors with perfectly overlapped categories.

2.1c. Weights. Data may be weighted to compensate for differential sampling or response rates. This is done by specifying a particular variable as a weight variable. Whenever statistics are based on weighted data, the unweighted number of cases is also shown.

2.2 Analysis of Subgroups

It is frequently useful to analyze separately data from certain subgroups. Whenever variables are expected to be related differently to one another in several subgroups (i.e., when there are interaction effects), this strategy may lead to a clearer understanding of what is happening. A second reason for analyzing subgroups separately might be to get some idea of the stability of the findings—are they approximately the same (for example) in two independent halves of the sample?

The MCA program has a powerful, optional "filtering" capability. A filter permits only cases with specified codes on specified variables to enter the analysis. When using the MCA program, it is possible to specify a global filter, which excludes cases from the complete MCA run, or local filters, which exclude cases from particular analyses, or both. (The capacity for multiple analyses within a single program run is discussed in the next section.) Filtering on a variable with randomly assigned codes provides a random subset of the data, but the analyst should realize that a random subset of a clustered sample is not truly independent of the rest of the sample. To get independent part-samples the sample design must be taken into account.

2.3 Multiple Analyses

Often a researcher knows that he will need several analyses. Sometimes he will want parallel analyses for different subgroups or analyses for different dependent variables. Other times he will want analyses which have nothing in common other than that they are all performed on the same set of individuals, or subgroups from this set. By using more than one "analysis packet" the researcher can do such multiple analyses within a single run of the program. When the researcher wants to do analyses which differ only in the set of predictors used he can use an even simpler option: he can supply multiple predictor lists within a single analysis packet. This option is convenient for determining the marginal contribution of a predictor or a set of predictors. One important use is for analysing data in which one set of explanatory variables can be thought of as prior to a second set.

2.3a Analysis packets. When running the MCA program, the user may supply as many "analysis packets" as he desires. Each analysis packet specifies a dependent variable, a group of cases (possibly all), parameters for the program, and one or more sets of predictors. For example, several packets, with different local filters, can be used if several analyses on slightly different subsets of the input data are to be made (e.g., when separate analyses are to be performed for males and females and for the total group). Or, to take a second example, two packets, with different parameter settings can be used if one analysis is to exclude cases with large values of the dependent variable and a second is to include them (in order to determine the effects of a skewed distribution on the analysis).

2.3b Multiple predictor lists within an analysis packet. If any particular analysis packet has more than one predictor list, a separate analysis will be done for each set of predictors; except for the predictors, the analyses will be identical. This program feature is particularly useful for determining the marginal or unique explanatory power of a predictor or a group of predictors. (Actually, there are three possible measures for assessing marginal predictive power. They are discussed, and procedures for computing them are given, in section 4.3. Two of them require adjusted R-squares from analyses which differ only in the predictors used.) A common reason for wanting to determine the marginal contribution of a set of predictors over and above a second set is parsimony. Another common reason is that one set can be thought of as logically prior to the other.

2.3c Strategy for models with logically prior sets of variables. Frequently our models of the real world are more complex than one dependent variable with an additive set of explanatory variables all of which are at the same causal stage. The most common complexity arises then the explanatory variables can be thought of as containing two sets, one of which is logically prior to the other and affects it or works through it. Suppose our variables were:

| Background variables (age, education, race, region) | Behavior and motivational variables (mobility, attitudes, expectations) | Current earnings |

We might want to ask whether there is anything in the behavior and attitudes that is not simply an expression of background in affecting earnings, i.e, we might want the marginal contribution of the second set of variables on top of the first set. Alternatively, or in addition, we might want to know whether there is anything in the background variables that operates directly rather than through our measured behaviors and attitudes that effects earnings. These are questions of marginal contribution and can be handled using multiple predictor lists.

2.4 Iterations and Convergence: User Options

The MCA model *implies* a set of normal equations. The computer program *uses* an iteration algorithm for approximating the coefficients constituting the solutions to these equations.[5] The iteration algorithm stops when the coefficients being generated are sufficiently accurate. This involves setting a tolerance and specifying a test for determining when that tolerance has been met. These are selected by the user. Three convergence tests are available. One requires only that the change in all coefficients from one iteration to the next be less than the criterion "Q" in absolute value. The parameter "Q" is supplied at the time of the analysis.

$$|a_i - a_{i+1}| \leqslant Q$$

The second test, which is the one usually specified, causes the iterations to cease when the change is below a specified fraction of the grand mean. The

---

[5]The limitations of the iterative method are discussed in section 3.3; the formulas for the iterative procedure are given in section 4.5.

fraction often specified is .001, requiring that all coefficients change less than one-tenth of 1 percent of the grand mean.

$$\left| a_i - a_{i+1} \right| \leqslant Q\overline{Y}$$

The third test requires that the change be less than a fraction of the ratio of the standard deviation of the dependent variable to its mean.

$$\left| a_i - a_{i+1} \right| \leqslant Q(\sigma/\overline{Y})$$

Accuracy should not be set unnecessarily high. Most data collected by survey techniques contain samplings and rounding errors. When the dependent variable is a 1-0 dummy and one is dealing with percentages, it might be insisted that the coefficients be accurate to .005. If the mean is 50 percent, then a reasonable convergence criterion might be test number two with $Q = .01$. If the dependent variable is a number, such an annual income, with a mean of about $6000, and the coefficients are desired accurate to the nearest dollar (no adjustment off by more than $0.50), then the convergence test might be 5/6000 or .00008.

It should be noted that convergence test two may cause trouble if the grand mean is close to zero or is slightly negative (as it may be if the analysis is being performed on the residuals from a previous analysis). In general, this will not be a source of trouble.

The user must also specify the maximum number of iterations the algorithm may try. If the coefficients do not converge before the specified number of iterations, the program will try one more and then print out its results on the basis of this last one.

The number of useful iterations depends somewhat on the number of predictors used in the analysis. Fewer predictors will converge sooner, if they are going to converge at all. If there are fewer than ten predictors it has usually been found satisfactory to specify ten as the maximum number of iterations permitted. Other analyses of more numerous and complex predictors have specified a maximum of as many as fifty iterations. The number of iterations required also depends on the fraction specified for tolerance.

Because the cost of an additional iteration is small compared with other costs, it is safest to set the number of iterations at a high level. This may mean paying for some extra iterations in the few instances when the coefficients do not converge (if they do converge, the number of iterations permitted is irrelevant). However, the cost of this is less than having the iterations stop prematurely. If the fraction specified for tolerance is set at .001 for convergence test two, twenty-five iterations are probably safe and adequate for most analyses.

2.5 One-way Analysis of Variance

If only one predictor is specified, the program does a conventional analysis of variance. This program capability, which does not involve the Multiple Classification Analysis technique, is provided for the convenience of the user; as described in the next chapter (section 3.2), a one-way analysis of variance can be used in conjunction with a MCA analysis to detect interactions among predictors. There are several exceptions associated with the one-way analysis of variance option: the number of predictor categories is 3000 rather than 32, codes must be in the range 0-2999, the various options for residuals are not available, and the option to supply more than one set of predictors within a single packet is not available. The printout which results from a one-way analysis is described in Appendix B.

2.6 Residuals

Optionally, the MCA program will compute, for each analysis, the predicted value for each individual and the deviation of his actual value from the predicted value. The residuals can be printed, written into a dataset, or both. If a subset of cases is analysed, the program can either produce residuals for just that subset, i.e., apply the MCA model to the same sample from which it was developed, or, alternatively, the program can produce residuals for a larger group by applying the model to cases in addition to those on which it was developed.

The residuals option is useful if, for example, one wishes to analyze residuals against other explanatory variables, to examine their distribution for normality, or to look at extreme residuals for evidence of conceptual or measurement errors or to stimulate ideas about other explanatory variables.

As noted above, if residuals are requested, *predicted* values for each case are computed. The predicted values can be used as a predictor or explanatory variable in a subsequent analysis as an "instrumental variable," i.e., an estimate of the original variable with much of its error removed. Errors in predictor variables in regression are very damaging, and such a two-stage procedure is often recommended in econometrics texts to reduce the bias in estimated regression coefficients.

3

Limitations of the Program

3.0 Summary

There are two principle sources of limitations in the program: the basic analytical model and the procedure used to solve the equations implied by the model. In addition, there is a theoretical limitation, that the sample size be large enough, and a practical limitation, that the size of the problem not exceed the allocated core space.

The first limitation arises because the method assumes that the data are understandable in terms of an additive model. An important implication of this is that the program is normally insensitive to interaction effects. Examples are presented showing data for which an additive model would be appropriate, and other data for which an additive model would not be appropriate because interaction is present.

If one knows interaction effects do not occur in one's data, there is no problem. If relatively simple interactions are known to be present, one may capture them by using new sets of categories, each defining an appropriate combination of two or more variables. If one is in doubt about the presence of interaction, there are several ways to explore the extent of its presence.

The second limitation arises from the iterative procedure the program uses to solve the normal equations required by the additive model. The procedure itself is described below. Various ways for checking the obtained coefficients are outlined. When categories are closely overlapped (predictors highly inter-correlated), the sequential adjustments do not converge rapidly, and estimates may still be changing when the interations are stopped. (In such a case conventional regression coefficients are also quite unstable.) If the overlap is too great, as when one subclass of each of two predictors is identical with the other, then the process does not converge at all. (In this case the solution to the usual normal equations of multiple regression becomes *impossisible*.)

The third limitation, that there be a large number of cases, arises from the large number of degrees of freedom associated with most MCA analyses and the fourth, having to do with problem size, from the program design.

3.1 Additive Models and Their Limitations

The program assumes the phenomena being examined can be understood in terms of an additive model. In other words, it assumes that the average score (on the dependent variable) for a set of individuals is predictable by *adding together* the "effects" of several predictors. An important implication of this is that the results can be distorted by interaction[1]. By a recoding technique described below, this limitation of the program sometimes can be overcome.

For readers who are unfamiliar with the concept of interaction, some examples may help to distinguish between the presence and absence of interaction.

Consider a dependent variable Y which has a mean of 10 for the full set of units one is examining. There are two predictors: A and B, and each has two categories, "1" and "2." In Figure 2, units in the first category of predictor A occur in the upper row, those in A=2 in the lower row, those in B=1 in the left column, and those in B=2 in the right column. Of course, units in the "1" categories of *both* A and B appear in the upper left cell. For each row, column, and cell there is a mean value of Y. If we make the simplifying assumption of equal numbers of cases in all the cells (i.e., if we assume predictors A and B are uncorrelated)[2] the mean value of Y for each row, column, and cell would be as shown in Figure 2. As shown there, the "effect" of being in category A=1 is to raise the score 2 units above the grand mean (note the deviation $\bar{Y}_{1.} - \bar{Y}..$). Similarly, the effect of being in B=1 is to raise the score 3 units above the grand mean. And if a unit is in both A=1 and B=1 its score is 5 (2+3) units above the grand mean. Thus an additive model which says the "predicted value for a cell mean is the sum of the grand mean plus the A effect plus the B effect" exactly fits the data.

Figure 3 (in which we also assume equal numbers of cases in all cells) presents a rather different situation. It shows an interaction effect. Neither predictor *by itself* has any "effect" on the dependent variable (note the zero deviations of the marginal means from the grand mean). Neverthless, the mean value of Y does vary under certain *combinations* of predictors (as shown by the means within the cells). Stated another way: the effect of one predictor on the dependent variable depends on the level of the other predictor. Interaction occurs in many forms, Figure 3 being one example. An additive model

---

[1] Alternative names for interaction include: conditioning effects, contingency effects, moderator effects, and specification effects.

[2] The program does not assume no intercorrelation among predictors; indeed its purpose is to deal with such intercorrelations. The only restriction is that they cannot be too high, or the estimates become unstable. For further discussion of the distinction between intercorrelated predictors and interaction effects, see Morgan and Sonquist (1963).

would not provide a good understanding of the data in Figure 3, nor does an additive model provide a good understanding of many data encountered in actual research. For example, much of social science theory contains hypotheses about interactions: the relationship of variable A to variable B depends on social class, relative deprivation, stage in the life cycle, etc. Thus interactions commonly occur and often are of considerable interest.

	B=1	B=2	
A=1	$\overline{Y}_{11} = 15$	$\overline{Y}_{12} = 9$	$\overline{Y}_{1.} = 12$ $(\overline{Y}_{1.} - \overline{Y}_{..} = +2)$
A=2	$\overline{Y}_{21} = 11$	$\overline{Y}_{22} = 5$	$\overline{Y}_{2.} = 8$ $(\overline{Y}_{2.} - \overline{Y}_{..} = -2)$
	$\overline{Y}_{.1} = 13$ $(\overline{Y}_{.1} - \overline{Y}_{..} = +3)$	$\overline{Y}_{.2} = 7$ $(\overline{Y}_{.2} - \overline{Y}_{..} = -3)$	$\overline{Y}_{..} = 10$

FIGURE 2
ADDITIVE EFFECTS

Shows fictitious data for which an additive model would be appropriate. By adding the "effects" of membership in certain categories to the grand mean, the corresponding cell mean is obtained. (It is assumed there are equal numbers of cases in all cells.)

	B=1	B=2	
A=1	$\overline{Y}_{11} = 12$	$\overline{Y}_{12} = 8$	$\overline{Y}_{1.} = 10$ $(\overline{Y}_{1.} - \overline{Y}_{..} = 0)$
A=2	$\overline{Y}_{21} = 8$	$\overline{Y}_{22} = 12$	$\overline{Y}_{2.} = 10$ $(\overline{Y}_{2.} - \overline{Y}_{..} = 0)$
	$\overline{Y}_{.1} = 10$ $(\overline{Y}_{.1} - \overline{Y}_{..} = 0)$	$\overline{Y}_{.2} = 10$ $(\overline{Y}_{.2} - \overline{Y}_{..} = 0)$	$\overline{Y}_{..} = 10$

FIGURE 3
INTERACTION EFFECT

Shows data for which an additive model would be completely inappropriate. Neither the A nor the B predictor by itself shows any "effect" on the dependent variable, but there is a marked AXB interaction. (All cell N's are assumed equal.)

It is possible to have both additive main effects as in Figure 2 *and* interaction effects as in Figure 3. Figure 4 shows how such data might look. Here an additive model partly explains the data but fails to capture the interaction. In this case there are two sources of "errors" in predicting the Y value for a given individual. The first is due to neglect of the A x B interaction effect which accounts for the departure of the observed cell mean from the predicted cell mean. The second is due to neglect of other factors which account for the departure of the individual's own score from the observed mean for his cell. (Although these two sources of errors are conceptually distinct, they are not separable in any one run of the MCA program. Chapter 5 further discusses the matter.)

	B=1	B=2	
A=1	$\overline{Y}_{11} = 17$	$\overline{Y}_{12} = 7$	$\overline{Y}_{1.} = 12$ $(\overline{Y}_{1.} - \overline{Y}_{..} = +2)$
A=2	$\overline{Y}_{21} = 9$	$\overline{Y}_{22} = 7$	$\overline{Y}_{2.} = 8$ $(\overline{Y}_{2.} - \overline{Y}_{..} = -2)$
	$\overline{Y}_{.1} = 13$ $(\overline{Y}_{.1} - \overline{Y}_{..} = +3)$	$\overline{Y}_{.2} = 7$ $(\overline{Y}_{.2} - \overline{Y}_{..} = -3)$	$\overline{Y}_{..} = 10$

FIGURE 4
ADDITIVE EFFECTS AND INTERACTION EFFECT
Shows how the data would look if the additive effects shown in Figure 2 were combines with the interaction effect shown in Figure 3.

3.2 What to Do About Interaction

3.2a Interaction is known to be absent. If one knows an additive model is appropriate, one can use the program without concern for its insensitivity to interaction.

3.2b Interaction known to be present. If one knows certain interactions occur, one can use a combined variable, sometimes called a pattern variable, to overcome this limitation. For example, a combination predictor might be created from age and education variables. Or several variables might be used to form a family life cycle predictor.

The technique is straightforward. Consider the data shown in Figure 3. Instead of entering two separate predictors, A and B, one would combine them into a new variable, AB. The new variable might have a separate category for each combination of categories in the original variables, thus:

Variable A	Variable B	Variable AB
1	1	1
1	2	2
2	1	3
2	2	4

(Or, of course, one might combine some of the possible categories on the AB variable into a single category.) The effects attributed by the program to the various categories of predictor AB would include both the main effects and interaction effects of variables A and B.[3] Note that original A and B variables should be omitted from the analysis.

This technique is frequently useful for including two-way interactions in the analysis. In principle it could be used to include three-way and even higher order interactions. But soon the total number of possible "interaction variables" thus generated becomes unreasonably large, being the product of the numbers of subclasses of each predictor: K_1 x K_2 x K_3 x K_4. . . . Either of two things may happen. The number of subclasses may get as large as the number of cases, and then the results can "explain" everything in the sample, but will not necessarily predict well to another sample. Or else many of the subclasses will be empty because no such combination exists.

Another possibility for handling interaction involves running separate analyses. If almost all of the interactions involve the same predictor, analyses of disjoint groups may be called for. A variation on this procedure is to run two analyses, one for the whole population and one for some subgroup of the population. If one merely wants to see whether the interaction biases the estimates for the whole population seriously, one can run the subgroup analysis with the group that makes up the largest part of the sample; if one wants to know whether there are different patterns of effects for some small subgroup, an analysis must be run for that small subgroup.

3.2c Presence of interaction in doubt. One advantage of creating all possible interactions is that it is possible to determine what the maximum degree of explanation would be, using any combinations of the basic predictors that did any good (Horst 1954; Lubin and Osborne 1957). The weighted sum of squares of the means of all possible subcells is the total possible explained

---

[3]A Multiple Classification Analysis must have at least two predictors. If the variable AB were the only predictor of interest, one could enter a constant as a second "predictor." This constant would permit the program to run, but would not otherwise affect the results. Or, one could enter only the predictor AB and do a one-way analysis of variance. Although the form of the printout would differ slightly depending on which approach was used, the substantive result would be the same.

variance, with the predictors at hand. If the variance explained by the MCA analysis assuming additive effects is substantially smaller, then there are substantial interaction effects. There is no other sure test.

Short of this, one might generate all the possible two-way subcells. The superiority of their explanatory power over the additive MCA results would indicate the potential power of first-order interaction effects.

Or one could run a MCA and examine the residuals. A set of residuals less normally distributed than the original dependent variable, though unlikely, might betoken some interaction effect. But large unexplained residuals or a nonnormal distribution of residuals may merely mean failure to introduce the proper explanatory variables, or to measure them well.

If, however, it is possible to select, on the basis of theory, experience or hunch, some explicit interaction effects which might be introduced, a special feature of the MCA program can be used to determine the extent of interaction. The feature is this: if only one predictor is specified, the program performs an ordinary one-way analysis of variance. Using a combination variable in a one-way analysis can assist a MCA user in detecting predictor interactions. The complete procedure is as follows:

—Determine a set of suspected interacting predictors.
—Form a "combination variable" using these predictors.
—Run one MCA analysis using the suspect predictors to get adjusted R^2.
—Run one MCA analysis with the "combination variable" as the control in one-way analysis of variance to get adjusted eta squared, which will be $\geqslant$ adjusted R .
—Use the difference, adjusted eta squared—adjusted R (the fraction of variance explained which is lost due to the additivity assumption), as a guide to determine whether the use of a combination variable in place of the original predictors is justified.

If, on the other hand, the problem is to locate the most important interaction effects it should be solved before the MCA program is run. One way to locate interactions is to use AID (Automatic Interaction Detector program), a computer program which was designed for the purpose (Sonquist, Baker, and Morgan 1973; Sonquist 1970).

3.3 The Iterative Procedure and Its Limitations

The central aspect of the program is its ability to determine the "coefficients" or "adjusted deviations" associated with the categories of each predictor. These adjusted deviations represent the program's attempt to fit an additive model to the input data. The values of these coefficients can be obtained by solving a set of simultaneous linear equations called the normal equations. The program actually arrives at these coefficients, however, by a series of successive approximations, altering one coefficient at a time on the basis of its latest estimates of the other coefficients.

It can be shown that the adjusted deviations produced by the program are mathematically identical to those which could be derived by solving the normal equations of ordinary regression, using "dummy variables" and adjusting the coefficients so that the weighted mean of each set was equal to zero. When intercorrelations among the predictors are too high, however, the iterative process used by the MCA program may fail to converge or converge only slowly with oscillations, while the ordinary regression will signal trouble by producing large standard errors. If two subclasses are actually identical in composition (perfect intercorrelation between two dummy variables) the MCA iteration algorithm will fail to converge, while an ordinary regression program will usually signal a matrix that cannot be inverted.

A comprehension of the iterative process is useful for an understanding of this program, and for providing an intuitive understanding of what either the multiple regression or MCA techniques do; i.e., how they estimate the effects of one predictor while "taking account" of the others. The process is closely analogous to Bean's method of graphic curvilinear multiple correlation (Bean 1929).[4]

3.3a The iterative process. One must remember that each category of each predictor has an adjusted deviation (coefficient) associated with it. One complete cycle or "iteration" in the process of successive approximations consists of a recalculation of each coefficient. Let us trace what happens to a given coefficient during the course of a single iteration.

Consider the example described in Chapter 1. In trying to account for variation in scientists' output of reports, we used three predictor variables: "highest degree," "professional experience," and "time in laboratory." One of the categories of the first predictor was "PhD" and there would be a coefficient associated with it. Recall, however, that the people who fell in the "PhD" category on this first predictor *also* fell in one or more categories on the second and third predictors. In other words, there were other coefficients which could also be associated with these people due to their classifications under the second and third predictors.

The recalculation of the coefficient for the "PhD" category during a single iteration would proceed as follows: (1) Compute the deviation of the mean for PhD's (on the dependent variable) from the grand mean; (2) Adjust this deviation on the basis of a weighted average of the latest estimates of coefficients from *other* predictors which also apply to the cases in this "PhD" category; (3) Store this adjusted deviation as the machine's latest estimate of the coefficient (deviation) for this category. Thus it can be seen that the recalculation of each coefficient is dependent upon a weighted average of certain other coefficients. If the subgroup is distributed on the other predictors

---

[4]See also Ezekiel and Fox (1959) Chapter 16. The formulas for the iterative process appear in Chapter 4.

exactly as on the whole sample, then the weighted sum of coefficients for each of the other predictors will be zero. (Where there is no intercorrelation among predictors, no adjustments are needed.)

At the end of each iteration, each coefficient has a value which is usually different from its value on the previous iteration. For any particular coefficient one can trace how it changed as the iterations progressed. During the early iterations these changes may be relatively large and may be either up or down. In later iterations the changes usually occur in just one direction and become smaller and smaller as the successive estimates converge on the "true" value. Figure 5 shows the values a coefficient might have as iterations progress.

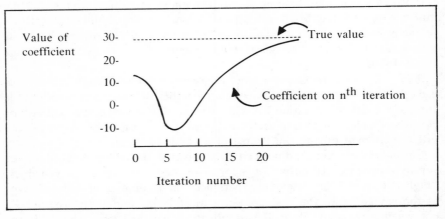

FIGURE 5
PROGRESS OF A COEFFICIENT TOWARD ITS CORRECT VALUE

How does the iteration procedure terminate? It would be convenient to test whether all coefficients had converged with their true values. But this, of course, is impossible since the true values are unknown. The "convergence test" actually used examines how much each coefficient changed from one iteration to the next. When a given coefficient changes by less than an amount specified by the user, it has "met the convergence test." When all coefficients have met the test, iterations stop.

Since it is uncertain how many iterations will be required before all

---

[5]If one or more coefficients have not converged by the time the maximum number of iterations is reached, the output will contain the values of each coefficient on the final two iterations. By examining the difference between these iterations (printed by the program), one can identify which coefficient or coefficients changed more than the convergence criterion.

coefficients meet their tests, the user is also required to specify a maximum number of iterations. If this maximum number is reached before all coefficients have converged, iterations stop anyway. Thus this is a second way, added as a safety precaution, for iterations to be stopped.

3.3b Situations which cause problems. As indicated, simultaneous estimates of coefficients either by matrix inversion (solution to the normal equations) or iteration, is not necessary if the predictors are all uncorrelated, and impossible if there is a perfect correlation among two or more predictor subclasses. The correlation may result from one being a perfect linear combination of *several* others. In practice, most problems seem to arise because a few categories are perfectly overlapped, as when the "retired" are put in a special "inappropriate" class for answers to a question about earnings. Figure 6 shows such a situation. In this case, a regression with dummy variables would be insoluble, while the iterative MCA program would not converge, and would tell the user which two coefficients were causing the trouble (in this case A-2 and B-1).

If the user discovers some categories are perfectly overlapped, he must either recode the data by combining categories or must omit the offending cases.

	B=1	B=2	B=3
A=1		X	X
A=2	X		
A=3		X	X

FIGURE 6
PERFECTLY OVERLAPPED CATEGORIES
Only the cells marked with X's contain cases. No units fall in the blank cells. Patterns such as these cannot be analyzed by the MCA program because some categories are perfectly overlapped (A=2 and B=1).

3.3c Closely overlapped categories. Even if categories are not perfectly overlapped, there may be only a small number of cases which prevent a condition of perfect overlap. In this case we shall speak of the categories as being "closely" overlapped.

When predictor categories are closely overlapped, coefficients may move rather slowly toward the values that constitute a solution of the normal equations. If iterations continued long enough, eventually the correct values might be achieved, but often iterations are stopped before this happens.

If the progress of a coefficient toward its correct value is very slow, it may change so little from one iteration to the next that it meets its convergence test prematurely. In this case, a too lenient test is being used.

On the other hand, slow progression of coefficients toward their correct values may result in the maximum number of permitted iterations being reached before all coefficients have met their tests. This will also stop iteration.

Two of the optional printouts from the program can assist in the identification and solution of these problems.

One optional printout is a set of cross tabulations for each predictor against each other predictor. From an inspection of these, the user can determine which categories, if any, are closely overlapped. If any such categories occur, their associated coefficients should be inspected with great care.

The second optional printout—the values of all coefficients on all iterations—can help to decide whether the suspect coefficients were close to their correct values when iterations stopped. By plotting their values at each iteration, one can see how they were changing. A rerun of the program allowing more iterations or requiring a more stringent convergence test or both may prove desirable.

Another way to check the obtained coefficients is to make a second run of the program entering the predictors in a different order. When certain categories are closely overlapped, it is often particularly enlightening to reverse the order of these two predictors. Since the iterative procedure makes successive approximations to each coefficient on the basis of the latest estimates of other coefficients, changing the order of predictors results in an entirely new solution as far as the program is concerned. If the coefficients obtained in the two runs are similar, the user's confidence in them is increased.

3.3d What to do when iterations fail to converge. For reasons described above it may occasionally happen that the iterative procedure will fail to converge. What should the user do?

First, he should determine which particular coefficients had not met the convergence test when iterations were terminated. In many cases it will turn out that there are only a few such coefficients and that these pertain to categories with very small numbers of cases. Often these will be categories of little substantive importance—e.g., missing data categories. In such a case, one can usually use the results obtained, without any need to re-run the analysis.

If it is determined that coefficients which failed to converge include ones of substantive interest, the next step is to determine why convergence did not occur. One condition which may have caused non-convergence, but which may not require rerunning the analysis, is the inadvertent use of a too-stringent convergence criterion or too-stringent convergence test. If investigation shows that all substantively important coefficients changed by less than some acceptably small amount during the final iterations, these coefficients can be accepted despite the indication that the procedure did not converge.

If neither of the above reasons account for the non-convergence—i.e., it cannot be explained on the basis of involving only unimportant coefficients or selection of an inappropriate test or criterion, then an additional run of the program will be required, after correcting the cause of the non-convergence. Sections 3.3b and 3.3c have described a variety of problems which can cause iteration failure. The re-run will involve one or more of the following: increasing the number of iterations allowed, omitting one or more variables, combining several highly correlated variables into one index, collapsing several categories into a single category, or perhaps omitting certain cases.

3.4 Number of Cases

Two considerations govern the minimum number of cases appropriate for use with the MCA program. First, if one wants to make inferences beyond his own data, each category of each predictor must have enough cases to provide reasonably stable estimates of means. However, including categories with very small numbers, as, for example, the "not ascertained" cases, does no damage to the other estimates. Second, there should be substantially more cases than there are degrees of freedom in the predictive model. (The degrees of freedom is equal to the sum of the number of categories for each of the predictors minus the total number of predictors.) Although the program will usually produce output when the above conditions are not met, the results are likely to be meaningless.

The distinction between $R^2_{unadjusted}$ and $R^2_{adjusted}$ shows the importance of the second condition and points to an easy method for determining, before a MCA is run, if the condition will be met. $R^2_{unadjusted}$ is the actual proportion of variance in the dependent variable explained by using the obtained coefficients in an additive model applied to the data cases actually used in that analysis. $R^2_{adjusted}$ (generally the more useful of the two statistics) is an estimate of how much variance the obtained coefficients would explain if used in an additive model applied to a different (but comparable) set of data cases—e.g. the population from which the sample actually analyzed was drawn. R^2_{adj} is calculated from R^2_{unadj} using an adjustment factor which corrects for capitalization on chance in fitting the model; the adjustment factor, which depends on the number of cases, categories, and predictors, can be determined *before* the MCA is run. The user can then apply this factor to an hypothesized R^2_{unadj} to determine the corresponding R^2_{adj} and decide if the amount of shrinkage is acceptable. If not, either the number of cases must be increased or the scope of the analysis (degrees of freedom) used reduced.

The following example shows how the calculation can be done. The formula for the adjustment factor is:

$$A = \frac{N - 1}{N + P - C - 1}$$

where

N = number of cases
P = number of predictors
C = number of categories (summed across all predictors.)

The formula for R^2_{adj} is:

$$R^2_{adj} = 1 - (1 - R^2_{unadj}) \; (A).$$

Suppose there are 300 cases and 5 predictors, each predictor with 7 categories. Then $A = 1.11$. If it is assumed that the predictors explain .40 of the variance in one sample, the R^2_{adj} will be .33, a decrement of 18%.

It turns out that for a moderately predictive system (R^2_{unadj} approximately equal to .4) a good rule of thumb is to have about 10 times as many cases as degrees of freedom used.[6] For a set of predictors which predict very well, fewer cases are needed for the same percent decrement, while for a set that predicts poorly many more cases are needed.

3.5 The Core Storage Limitation

In any particular MCA analysis (or group of analyses within an analysis packet), it is not possible to use the maximum number of predictors, each with the maximum number of categories. To avoid core storage problems, the following formula should be used to estimate the number of predictors which can be used in an analysis packet:

$$4 \left(\sum_{i=1}^{P} C_i + \left(\sum_{i=1}^{P} C_i \right)^2 - \sum_{i=1}^{P} C_i^2 \right) / 2 \leqslant 12{,}000$$

where

C_i = maximum number of distinct codes for the ith predictor variable.
P = number of predictor variables.

If a problem exceeds the available core, an error message is printed and the program skips to the next analysis packet.

---

[6]Degrees of freedom = $C - P$.

Part Two

A Statistical Description of the Program

4

Formulas

4.0 Summary

This chapter gives the definition of the summary statistics printed by the program and information on how to compute certain statistics which are not printed. It also describes the underlying model of the program and gives the formulas for the iterative procedure.

4.1 Formulas for Statistics Printed by the Program

Adjusted eta^2 and F are not printed if the program is used for its primary purpose, Multiple Classification Analysis. Beta and the various forms of the multiple correlation coefficient are not printed if the program is used for one-way analysis of variance.

Basic Terms

Y_k = individual k's score on the dependent variable
w_k = individual k's weight
N = number of individuals
C = total number of categories across all predictors
c_i = total number of categories in predictor i
P = number of predictors
a_{ij} = adjusted deviation of jth category of predictor i on final iteration

Sum of Y	$= \sum_k w_k Y_k$	(4-1)
Sum of Y^2	$= \sum_k w_k(Y^2{}_k)$	(4-2)
Grand Mean of Y	$= \dfrac{\sum_k w_k Y_k}{\sum_k w_k}$	(4-3)
Sum of Y for category j of predictor i	$= \sum_k w_{ijk} Y_{ijk}$	(4-4)

Sum of Y^2 for category j of predictor i $= \sum_k w_{ijk} Y^2_{ijk}$ (4-5)

Standard deviation of Y $= \sqrt{\dfrac{\sum_k w_k Y_k^2 - (\sum_k w_k Y_k)^2 / \sum_k w_k}{\sum_k w_k - (\sum_k w_k / N)}}$ (4-6)

Mean Y for category j of predictor i $= \overline{Y}_{ij} = \dfrac{\sum_k w_{ijk} Y_{ijk}}{\sum_k w_{ijk}}$ (4-7)

Sum of squares based on unadjusted deviations for predictor i[1] $= U_i = \sum_j (\sum_k w_{ijk})(\overline{Y}_{ij} - \overline{Y})^2$ (4-8)

Sum of squares based on adjusted deviations for predictor i $= D_i = \sum_j (\sum_k w_{ijk})(a_{ij})^2$ (4-9)

Explained sum of squares[2] $= E = \sum_i \sum_j a_{ij}(\sum_k w_{ijk} Y_{ijk})$ (4-10)

$= \sum_i \sum_j \ldots \sum_n w_{ij\ldots n}(a_i + b_j \ldots)^2$ (4-11)

Total sum of squares $= T = \sum_k w_k (Y_k - \overline{Y})^2$ (4-12)

$= \sum_k w_k (Y_k^2) - \dfrac{(\sum_k w_k Y_k)^2}{\sum_k w_k}$

Residual sum of squares $= Z = T - E$ (4-13)

Eta for predictor i $= \eta_i = \sqrt{U_i / T}$ (4-14)

Beta for predictor i $= \beta_i = \sqrt{D_i / T}$ (4-15)

Multiple correlation coefficient (squared) $= R^2 = E/T$ (4-16)

Adjustment for degrees of freedom $= A = \dfrac{N-1}{N + P - C - 1}$ (4-17)

---

[1]This is identical with the between groups sum of squares if a one-way analysis of variance were applied to predictor i alone. See Chapter 5.

[2]For the second of these formulas, 4-11, see section 4.4 for an explanation of the notation.

Multiple correlation coefficient (squared and adjusted for degrees of freedom)[3]

$$=R^2_{adj} = 1 - \left[\frac{(T-E)/(N-C+P-1)}{T/(N-1)}\right] \quad (4\text{-}18)$$

$$= 1 - (1 - R^2)A \quad (4\text{-}19)$$

Eta (squared and adjusted[4] for degrees of freedom)

$$=eta^2_{adj} = 1 - \left[\frac{(T-E)/(N-C)}{T/(N-1)}\right] \quad (4\text{-}20)$$

$$= 1 - (1 - eta^2)A \quad (4\text{-}21)$$

$$F = \frac{E/(C-1)}{(T-E)/(N-C)} \quad (4\text{-}22)$$

The sampling stabilities of the etas, betas, unadjusted and adjusted coefficients are discussed in section 6.6.

4.2 Formulas for Two F Statistics Not Calculated by the Program

A variety of F tests can be computed from statistics printed by the program. It should be stressed, however, that the degree to which the data meet the assumptions of an F test is not determinable from the output. As always, the user bears the responsibility for interpreting the test.[5] Two F tests are described below. The first is a test for all predictors simultaneously; the second is a test for a single predictor taken by itself.

Test number one is a test for all predictors simultaneously. The "adjusted multiple correlation coefficient" printed by the program indicates *how much* of the variance of the dependent variable is accounted for by all the predictors together. The following F test answers the question: do all predictors together explain a significant portion of the variance of the dependent variable:

---

[3]If the value for the adjusted coefficient comes out a minus quantity, then 0 is printed.

[4]See note 3 above.

[5]When using weighted data, the interpretation of the F statistic becomes difficult. The formulas above are not applicable without introducing a correction for the effect of the weights on the sums of squares, and no ideal correction is available. Weights may be "normalized" so they sum to N. One rough approximation might be to divide each sum of squares term by the mean weight (the degrees of freedom terms, of course, use N rather than sum of weights; see formula (4-6) for the standard deviation formulas used. However, the user of even this approximation is warned that the assumption of independent random sampling, which underlies the F test, is being violated when the data are "weighted up." Furthermore, if the weights themselves bear any relationship to either the dependent or predictor variables, this may affect the value of the F statistic. There appears to be no easy general solution to this problem.

$$F = \frac{E / (C\text{-}P)}{Z / (N\text{-}C\text{+}P\text{-}1)} \cdot \qquad\qquad (4\text{-}23)$$

Test number two answers the question: does this predictor *all by itself* explain a significant portion of the variance of the dependent variable? This is the question answered by a classical one-way analysis of variance. An F test for predictor i which answers this question is:

$$F_i = \frac{U_i / (c_i\text{-}1)}{(T\text{-}U_i) / (N\text{-}c_i)} \cdot \qquad\qquad (4\text{-}24)$$

4.3 Procedures for Computing Measures of a Predictor's Importance

Several different criteria exist for assessing the importance of a predictor, i.e., the degree of relationship between an independent variable and the dependent variable or its "predictive power."

The eta statistic—defined in section 4.1—is appropriate for assessing the simple bivariate relationship between the predictor and the dependent variable. (Eta squared, sometimes called the correlation ratio, is interpretable as the proportion of variance in the dependent variable explainable by the predictor.)

The beta statistic—defined in section 4.1, and discussed in detail in section 6.3—is useful as an approximate measure of the relationship between a predictor and the dependent variable while *holding constant* all other predictors, i.e. while assuming that in each category of the predictor in question all other predictors are distributed as they are in the population at large.

Still a third criterion focuses on the "marginal" or "unique" explanatory power a predictor has over and above what can be explained by some specified set of other predictors. There are actually three different possibilities: [6]

1. One can remove the effects of the other predictors from the dependent variable, and then correlate the residuals (actual minus expected value of the dependent variable for each case) with the predictor in question. This produces one of two possible *part correlations*. As we have said in Chapter 1, this two-step procedure introduces a downward bias in both the regression coefficients and the second correlation coefficient, in proportion to the amount of correlation between the predictor in question and the other predictors (Morgan 1958; Goldberger and Jochem 1961).

2. One can remove the effects of the other predictors from the predictor in question, and then correlate the residuals of that predictor (actual minus

---

[6]For a more detailed discussion of the possibilities see Nunnally (1963, page 151-155).

predicted values) with the dependent variable. This is a different *part correlation*. It asks whether there is any variability in X not predictable from the other predictors that helps explain Y. In other words, this part correlation (when squared) assesses the importance of a predictor in terms of the variance in the dependent variable marginally explainable by the predictor, *relative to the total variance in the dependent variable.*[7]

One does not usually compute this part correlation by first getting the residuals. It can most easily be obtained by doing two MCA analyses, with and without the predictor in question, since the squared part correlation is equal to the increase in multiple R-squared:

$$\text{Squared part correlation} = (R^2_{adj} \text{ with everything in}) - (R^2_{adj} \text{ omitting one set}) \quad (4\text{-}25)$$

3. One can remove the effects of the other predictors from both the dependent variable and the predictor in question, and correlate the two sets of residuals. This is the partial correlation coefficient. It is the fraction of the remaining distance to 1.00 that one goes by adding the predictor in question. In other words, the partial correlation (when squared) assesses the importance of a predictor in terms of the variance in the dependent variable marginally explainable by the predictor relative to the *as-yet-unexplained variance.* As with the second of the part correlations above, one does not usually obtain it by actually performing the residualizations, but rather by two MCA analyses, with and without the predictor in question.[8] The squared partial correlation is estimable from the two multiple R-squares:

$$\text{Squared partial correlation} = \frac{(R^2_{adj} \text{ with everything in}) - (R^2_{adj} \text{ omitting one set})}{1 - (R^2_{adj} \text{ omitting one set})} \quad (4\text{-}26)$$

Within any one set of predictors, the second of the two part correlations discussed above will rank order the importance of predictors in a manner

---

[7] What we here denote as the squared part correlation is identical to what some writers would call the squared semi-partial correlation and to what Darlington (1968) has called the "usefulness" measure.

[8] In ordinary numerical regression, where each predictor variable has a coefficient and a standard error of that coefficient, the partial correlation squared has a simple relation to those two numbers and does not require running two regressions:

$$\text{Squared partial correlation} = \frac{(B/\text{standard error of B})^2}{(B/\text{standard error of B})^2 + N\text{-}K\text{-}1} = \frac{F}{F + df} = \frac{t^2}{t^2 + df}$$

identical to the rank order obtained from partial correlations. However, the numerical values of these two kinds of correlations are not the same, and they are not monotonically related across predictors which come from different sets.

While it is often true that beta coefficients and partial (or the second type of part) correlations rank order the importance of predictors within a set similarly, the ranking may not be identical (Ezekiel and Fox, 1959, p. 197). The relationships between beta and the partial correlation coefficient are discussed in detail in section 6.3.

4.4 The Statistical Model

The statistical model which the program uses is neither new nor complex. The model specifies that a coefficient be assigned to each category of each predictor, and that each person's (or other unit's) score on the dependent variable be treated as a sum of: the coefficients assigned to categories characterizing him, plus the average for all cases, plus an error term:

$$Y_{ij...n} = \overline{Y} + a_i + b_j + ... + e_{ij...n} \tag{4-27}$$

where

$$Y_{ij...n} = \text{The score (on the dependent variable) of}$$
individual n who falls in category i of
predictor A, category j of predictor B, etc.
$$\overline{Y} \quad = \text{grand mean on the dependent variable.}$$
$$a_i \quad = \text{the "effect" of membership in the } i^{th} \text{ category}$$
of predictor A.
$$b_j \quad = \text{the "effect" of membership in the } j^{th} \text{ category}$$
of predictor B.
$$e_{ij...n} = \text{error term for this individual.}$$

The adjusted coefficients can be thought of as having been estimated in such a way that they provide the best possible fit to the observed data, i.e., so as to minimize the sum of the (squared) errors. That set of coefficients can be obtained by solving a set of equations known as the normal equations (sometimes known as the least squares equations).

These equations received attention from Yates (1934, p. 60) in his "method of fitting constants." Subsequently, various other statisticians have described them including Snedecor (1946, p. 196), Kempthorne (1952, p. 91), and Anderson and Bancroft (1952, p. 279).[9]

---

[9]Anderson and Bancroft called the technique the "method of least squares."

All these writers fit constants which correspond closely to the coefficients printed out by the MCA program. Because Yates, Snedecor, and Kempthorne all make the assumption that the sum of the coefficients (unweighted for number of cases) should equal zero for each predictor, they also fit a constant for the mean. Anderson and Bancroft, however, make the assumption that the sum of the *weighted* coefficients should equal zero for each predictor. This implies that a mean, if it were fitted, would equal the observed mean. The normal equations and resulting coefficients will differ somewhat depending on which assumption is made, though the predicted scores for each individual will be the same. The MCA program follows Anderson and Bancroft in this respect.

The normal equations used by the MCA program are as follows (shown here for three predictors):

$$a_i = A_i - \overline{Y} - \frac{1}{W_i} \sum_j W_{ij}b_j - \frac{1}{W_i} \sum_k W_{ik}c_k$$

$$b_j = B_j - \overline{Y} - \frac{1}{W_j} \sum_i W_{ij}a_i - \frac{1}{W_j} \sum_k W_{jk}c_k$$

$$c_k = C_k - \overline{Y} - \frac{1}{W_k} \sum_i W_{ik}a_i - \frac{1}{W_k} \sum_j W_{jk}b_j$$

Where:

A_i = mean value of Y for cases falling in the ith category of predictor A.

B_j = mean value of Y for cases falling in the jth category of predictor B.

C_k = mean value of Y for cases falling in the kth category of predictor C.

W = number of cases (weighted).

4.5 Formulas for the Iterative Procedure

The program attempts to find values for the coefficients which will solve the normal equations by the following iterative procedure.[10,11]

---

[10] All terms are interpreted as previously defined for the statistical model or normal equations.

[11] An alternative approach to solving the normal equations would be to use a matrix inversion technique. After some experimentation, however, in which we actually tried matrix inversion, we concluded that the present iterative technique produces results which are as accurate as those obtainable from the matrix inversion routine we used, and required less computing time to do so. The savings in computing time increased as the size of the analysis increased and became very substantial (e.g., 40%) with even moderately sized problems (e.g., 5 predictors with 10 categories each).

Step (1) $a_i' = A_i - \overline{Y}$; $b_j' = B_j - \overline{Y}$; $c_k' = C_k - \overline{Y}$

Step (2) $a_i'' = a_i' - \dfrac{1}{W_i} \sum\limits_j W_{ij} b_j' - \dfrac{1}{W_i} \sum\limits_k W_{ik} c_k'$

$b_j'' = b_j' - \dfrac{1}{W_j} \sum\limits_i W_{ij} a_i'' - \dfrac{1}{W_j} \sum\limits_k W_{jk} c_k'$

$c_k'' = c_k' - \dfrac{1}{W_k} \sum\limits_i W_{ik} a_i'' - \dfrac{1}{W_k} \sum\limits_j W_{jk} b_j''$

Step (R) $a_i^R = a_i' - \dfrac{1}{W_i} \sum\limits_j W_{ij} b_j^{R-1} - \dfrac{1}{W_{ik}} \sum\limits_k W_i \sum C_k^{R-1}$ (Note: Superscripts denote primes, not powers.)

$b_j^R = b_j' - \dfrac{1}{W_j} \sum\limits_i W_{ij} a_i^R - \dfrac{1}{W_j} \sum\limits_k W_j \sum c_k^{R-1}$

$c_k^R = c_k' - \dfrac{1}{W_k} \sum\limits_i W_{ik} a_i^R - \dfrac{1}{W_k} \sum\limits_j W_j \sum b_j^R$

Step (R+1)

$a_i^{R+1} = a_i' - \dfrac{1}{W_i} \sum\limits_j W_{ij} b_j^R - \dfrac{1}{W_i} \sum\limits_k W_{ik} c_k^R$

$b_j^{R+1} = b_j' - \dfrac{1}{W_j} \sum\limits_i W_{ij} a_i^{R+1} - \dfrac{1}{W_j} \sum\limits_k W_{jk} c_k^R$

$c_k^{R+1} = c_k' - \dfrac{1}{W_k} \sum\limits_i W_{ik} a_i^{R+1} - \dfrac{1}{W_k} \sum\limits_j W_{jk} b_k^{R+1}$

5

Relations Between the MCA Technique and Analysis of Variance

5.0 Summary

Chapter 1 mentioned that the MCA program could appropriately be applied to data which did not meet the assumptions of the simpler, commonly used forms of analysis of variance. Nevertheless, the program is directly related to analysis of variance in its more complex forms. This was briefly mentioned in the discussion in Chapter 4 of the statistical model on which the program operates. It is the purpose of this chapter to make this relationship more explicit.

5.1 A Comparison of MCA with Complex Least-Squares Analysis of Variance

There are many extensive discussions of the more complex forms of analysis of variance. The authorities cited in Chapter 4—Yates, Snedecor, Kempthorne, and Anderson and Bancroft—are all relevant for the discussion which follows. We will adopt the terminology used by Snedecor, though not necessarily his notation or his assumptions.

Analysis of variance in its simpler forms usually assumes an equal number of cases in each of the "cells" formed by the cross classification of two or more predictors. For many years, however, statisticians have known how to perform more complex analyses of variance on data characterized by unequal numbers in the cells. In one sense, the MCA program is a computerized version of these long-known techniques for analysis of variance. Although it does not, when used for multivariate analysis, actually compute F tests, it does generate the various sums of squares terms from which a variety of F tests could easily be obtained. (Two of these F tests are described in Chapter 4.)

This chapter describes analysis of variance procedures which are appropriate when the full range of "complications" appear in the data, i.e. when there are unequal numbers of cases in the cells and when the presence of interaction is unknown. A two-predictor fixed-constants model is discussed, although the technique can be generalized to fixed-constants models with more predictors.

Figure 7 shows how data might be displayed in matrix format and indicates the notation used in this chapter.

	B_1	B_2	B_3		where:
A_1	N_{11} Y_{11} Y^2_{11}	N_{12} Y_{12} Y^2_{12}	N_{13} Y_{13} Y^2_{13}	$N_{1.}$ $Y_{1.}$ $Y^2_{1.}$	A,B = predictor variables Y = sum of the dependent variable for the group indicated Y^2 = sum of squares of the dependent variable for the group indicated
A_2	N_{21} Y_{21} Y^2_{21}	N_{22} Y_{22} Y^2_{22}	N_{23} Y_{23} Y^2_{23}	$N_{2.}$ $Y_{2.}$ $Y^2_{2.}$	N = number of cases for the group indicated
	$N_{.1}$ $Y_{.1}$ $Y^2_{.1}$	$N_{.2}$ $Y_{.2}$ $Y^2_{.2}$	$N_{.3}$ $Y_{.3}$ $Y^2_{.3}$	$N_{..}$ $Y_{..}$ $Y_{..}$	First subscript = row or A value Second subscript = column or B value Dot subscript = total over that dimension

FIGURE 7
ANALYSIS OF VARIANCE NOTATION
Shows a general matrix format and accompanying notation for an analysis of variance with two predictors and one dependent variable.

The first step in the analysis of variance is to determine the "true" effects of the A and B categories. This is accomplished by solving simultaneously a set of equations. These are the normal equations (shown in Chapter 4), or are a set of equations derivable from them. The resulting "fitted constants" (one for each category of each predictor) are identical with the "coefficients" or "adjusted deviations" computed by the MCA program.[1] These provide the desired indication of the "true" effects of the predictors on the dependent variable.[2]

Using these fitted constants, or coefficients, the analysis of variance then proceeds by calculating various sums of squares terms as follows.

Total sums of squares (of deviations about the grand mean)
$$= Y^2_{..} - \frac{(Y_{..})^2}{N_{..}}$$

Sum of squares attributable to subclasses
$$= \sum_i \sum_j \left(\frac{(Y_{ij})^2}{N_{ij}}\right) - \frac{(Y_{..})^2}{N_{..}}$$

Sum of squares within cells
$$= \text{(Total sum of squares)} - \text{(Subclasses sum of squares)}$$

Reduction in sum of squares due to fitting constants[3]
$$= \sum_i (a_i \, Y_{i.}) + \sum_j (b_j \, Y_{.j})$$

Where: a_i and b_j are the constants previously fitted to the several categories of the predictors.

Interaction sum of squares
$$= \text{(Subclasses sum of squares)} - \text{(Reduction in sum of squares due to fitting constants)}$$

---

[1] This statement is true only within limits of rounding, and holds only when the iterative procedure used by the program gives the intended results—see Chapter 3 for a discussion of conditions which may prevent this.

[2] As discussed in Chapter 4, statisticians make one of two assumptions with respect to these fitted constants. One assumption says that the sum of the weighted constants for any predictor should sum to zero; the other says they should sum to zero without weighting. Under the latter assumption one must fit a constant for the overall mean as well as for each category of each predictor. Under the former assumption the fitted mean would of necessity equal the observed mean and hence does not need to be fitted. The normal equations and resulting constants will differ somewhat depending on which assumption is made, although the predicted scores for each individual will be the same. The MCA program uses the former assumption.

[3] This formula assumes that constants were fitted under the constraint that $0 = \sum_i N_{i.} a_i = \sum_j N_{.j} b_j$, i.e. that the weighted sum of the constants is zero. If the alternative assumption that the unweighted sum is zero is made, a slightly different form of this formula must be used. Both assumptions provide the same reduction in sum of squares when the appropriate formulas are used.

If an F test applied to the obtained interaction sum of squares proves to be significant, the usefulness of going on to examine the separate effects of the predictors is debatable. (See Snedecor, 1946, p. 300, for further discussion and appropriate methods.) If, however, interaction is negligible, one can then proceed to obtain sums of squares attributable to the A and B effects by the following.

Gross sum of squares attributable to predictor A
$$= \sum_i \frac{(\Sigma Y_{i.})^2}{N_{i.}} - \frac{(Y..)^2}{N..}$$

Gross sum of squares attributable to predictor B
$$= \sum_j \frac{(\Sigma Y_{.j})^2}{N.j} - \frac{(Y..)^2}{N..}$$

Correction to gross sums of squares for A and B
= (Total of gross sums of squares attributed to A and B) - (Reduction in sum of squares due to fitting constants)

Net sum of squares attributable to predictor A (over and above what B could explain)
= (Gross sums of squares for A) - (Correction)

Net sum of squares attributable to predictor B (over and above what A could explain)
= (Gross sums of squares for B) - (Correction)

Thus the MCA program is closely linked conceptually to classical techniques for complex analyses of variance. The "Coefficients" approximated by the MCA program's iterative procedure are identical to the classical "fitted constants." And the MCA Program's "explained sum of squares" is identical to the classical "reduction in sum of squares due to fitting constants." What the program calls "residual sum of squares," is the sum of the "within sum of squares," plus "interaction sum of squares," if any. These various relationships may perhaps be best seen in terms of Figure 8. This figure shows how the total sum of squares would be "divided up" by analysis of variance and the MCA program.

Faced with the relationships in Figure 8, it is tempting to assume that the sums of squares based on adjusted deviations (obtained by the program), when added together, should equal the explained sum of squares. While this is the case when predictors are orthogonal, it is not generally true. With correlated predictors, the sum of these sums of squares may be either greater or less than the explained sum of squares. Similarly, the sum of the squared betas may be either more or less than the total proportion of variance explained; usually more. See Chapter 6 for further discussion of these points.

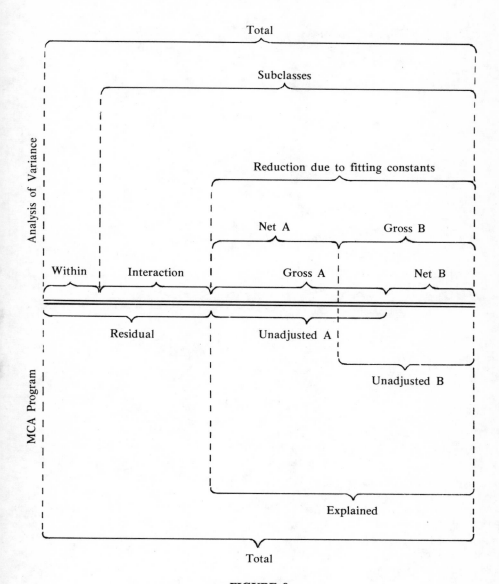

FIGURE 8
DIVISION OF TOTAL SUM OF SQUARES
Shows how the total sum of squares is "divided up" by analysis of variance and the MCA program, assuming two correlated predictors.

6

Relations Between the MCA Program and Multiple Regression

6.0 Summary

The previous chapter showed how the MCA program could be considered as a computerized version of a complex analysis of variance. The program can also be viewed as a form of multiple regression.

This chapter opens with a short description of dummy variable multiple regression. It then considers similarities between the coefficients resulting from dummy variable multiple regression and the MCA program. Next, the interpretation and derivation of a partial beta coefficient is discussed. The chapter closes with brief mention of other multiple regression techniques.

6.1 Multiple Regression Using Dummy Variables

Multiple regression with dummy variables is a technique which allows use of the usual multiple regression equations when predictors are nominal scale classifications. When predictors are measured on ordinal or interval scales, however, dummy variable multiple regression may still be useful because it requires no assumptions about relationships being linear. Thus this is a flexible technique useful for a wide range of problems.

Each "dummy variable" tells whether a person (or other unit) falls within a particular classification. In terms of the example discussed in Chapter 1, the predictor variable "highest degree," would yield a set of three dummy variables, each scored yes or no (1 or 0) whether the highest degree was: (a) a BS, (b) an MA, or (c) a PhD. Similarly, there would be a set of four dummy variables derived from the predictor variable of professional experience and five from time in laboratory. In all, then, 12 (3+4+5) dummy variables might be generated for the example described in Chapter 1.

It happens, however, that inclusion of all these dummy variables would make this problem unsolvable. The multiple regression equation would have perfect correlation among the predictors. If a man's membership in each class but one is known, his membership on the last class is perfectly derivable. To avoid such "over-determination," one of the coefficients is usually constrained to zero for each set of dummy variables—accomplished simply by omitting one of the dummy variables of each set from the analysis. Under this constraint, the obtained coefficients for the remaining dummy variables within each set indicate deviations *from the omitted variable*. In terms of our example, if the dummy variable "highest degree is BS" were omitted, coefficients obtained for the dummy variables would indicate how much the performance of scientists with Masters degrees tended to exceed that of scientists with Bachelors degrees. Suits (1957) provides a full discussion of these points.

6.2 Similarities in Coefficients

The coefficients obtained by the MCA program are analogous to those obtained by multiple regression using dummy variables. The conversion of any set of coefficients from one technique to those from the other is straightforward. The relationship is:

$$a_{ij} = b_{ij} + Q_i$$

Where: a_{ij} = MCA program coefficient for the jth category of predictor i.

b_{ij} = dummy variable multiple regression coefficient for the corresponding classification.

Q_i = an adjusting constant, described next.

Q is an adjusting constant which makes the weighted sum of the a_i's $= 0$ (the constraint used by the MCA program). There will be a different Q for each MCA predictor or set of dummy variables. The value of Q is:

$$Q_i = - \sum_j P_{ij}b_{ij}$$

Where: P_{ij} = the proportion of total cases falling in the jth category of the ith predictor.

The sum of the constant term from the multiple regression plus all the Q's should equal the grand mean of the dependent variables. This provides a check on the computation of the Q's.[1]

An example will help make this conversion clear. Let us again consider the data relating highest degree to output of scientific reports (expressed as percentile scores). The equivalences would be as follows if the coefficient for the BS group were constrained to zero (Q = - 6.5).

---

[1]Sweeney and Ulveling (1972) have extended this method of transformation to conventionally scaled independent variables when used in combination with dummy variables.

Group	Proportion of Sample	Dummy Variable Regression Coefficients	MCA Coefficients
BS	.71	$b_1 = 0$ (omitted)	$a_1 = -6.5$
MS	.22	$b_2 = 16.5$	$a_2 = 10.0$
PhD	.07	$b_3 = 41.5$	$a_3 = 35.0$

Thus the MCA technique can be considered the equivalent of a multiple regression using dummy variables. However, MCA is often more convenient to use than multiple regression with dummy variables. These is no need to recode the predictor variables into sets of dummy variables prior to making an analysis. And the coefficients for *all* categories are obtained and expressed as deviations from the mean, a form in which they are easily understood.[2]

6.3 *The Partial Beta Coefficients: Interpretation of Beta*

Included in the output of the MCA computer program is a partial beta coefficient for each predictor. The rank order of these betas indicates the *relative* importance of the various predictors in their explanation of the dependent variable if all other predictors were "held constant."[3] To assess the *marginal* importance of a particular predictor two MCA analyses are necessary; various measures of marginal importance are discussed in section 4.3 and procedures for computing them are given.

To reiterate, the beta coefficients must be interpreted with caution and are useful only for indicating the relative importance of the various predictors. By the formulas given in Chapter 4, one can see that the square of the beta coefficient is the sum of squares attributable to the predictor (after "holding other predictors constant") relative to the total sum of squares. Thus it is tempting to interpret this beta coefficient (when squared) in terms of "per-cent of variance explained." This would be appropriate, however, *only* in the uninteresting special case when all predictors are uncorrelated and multivariate analysis is unnecessary. In general, one must not attempt to interpret the beta coefficient in this way. (The sum of squares based on adjusted devia-tions, when summed across all predictors, may add to either more or less

---

[2] As noted above, in dummy variable multiple regression one category of each predictor (i.e. one variable from each set of variables) is omitted, and coefficients for the remaining categories are expressed as deviations from the omitted coefficient. If the number of cases in the omitted category is small, its coefficient may be quite far from the overall mean, and hence all other coefficients may appear unusually large (relative to their standard errors). On the other hand, if a large group is excluded, the reader will have no information on its coefficient unless he performs additional calculations.

[3] See section 4.3 for a brief discussion of three distinct criteria which can be used to assess the importance of predictors. Betas are relevant to only one of these.

than the explained sum of squares, and can even *exceed* the total sum of squares. Hence, if one were to square each of the beta coefficients and sum them across the predictors, their sum might be more or less than the square of the multiple correlation coefficient, and can even be *more* than 1.00.)

It would be nice if one could "divide up" the total explanatory ability of several predictors and allocate each its "fair" portion.[4] But it can be argued that it does not make sense. If two correlated predictors each explain a portion of the variance in a dependent variable, there is little reason to divide up the portion of the variance that either can explain and speak of one explaining one part and the other explaining another part. It is simply true that *either* predictor can do equally well for this part of the variance.

6.4   Another View of MCA Betas

The beta used by the MCA program may be understood better by thinking of the following process, though in practice there is no need actually to carry out the computations.

Suppose one considers the coefficients from the MCA program (the a_{ij}'s) as numbers. There is, then, a set of numbers for each predictor. Each individual is assigned exactly one number for each predictor, depending on the class in which he falls. If we used these numbers as a new set of variables, a rescaling of each predictor into numbers, and ran a conventional multiple regression with the same dependent variable, it can be shown that the regression coefficient b_i for each of these variables would be equal to 1.0 and the constant term of this regression equation would be the mean of the dependent variable.

The formula for the conventional beta coefficient is:

$$\beta_i = \frac{b_i\, \sigma_{xi}}{\sigma_y}$$

But with this set of new variables, b_i is always equal to 1.0, the standard deviation of the dependent variable is known, and the standard deviation of our artificial variable is the weighted sum of squares of the coefficients which became the variable (because their mean is zero), divided by the sum of the weights. So the beta coefficient for a predictor used in the MCA program, algebraically equivalent to that shown in Chapter 4, is:

$$\beta_i = \frac{1}{\sigma_y}\sqrt{\sum_j N_{ij} a_{ij}^2 / N}$$

Where: N_{ij} = number of cases in the jth category of predictor i.

---

[4]Figure 1 in Chapter 1 may help in understanding this discussion.

Since the coefficients obtained from the MCA program can be converted into those obtained from a multiple regression using dummy variables (and vice versa), we can convert the above expression into a beta appropriate for the coefficients as they come from multiple regression techniques with one class excluded.

The formula becomes:

$$\beta_i = \frac{1}{\sigma_y} \sqrt{\frac{\sum_j N_{ij} b_{ij}^2}{\sum_j N_{ij}} \left(\frac{\sum_j N_{ij} b_{ij}}{\sum_j N_{ij}}\right)^2}$$

The formulas differ because the a_{ij} have a weighted mean of 0 whereas the b_{ij} do not.

Since the "variable" involved is more flexible than the usual numerical variable used in conventional multiple regression, each coefficient being given its appropriate sign, no sign is attached to these beta coefficients. In all other respects, however, they appear analogous to the usual beta coefficients, and are subject to the same advantages and disadvantages.

It is interesting, since these betas are like the partial betas of regression with ordinary scaled variables, to note that there is a relationship between the partial beta coefficient of multiple regression and the true partial correlation squared. The relationship between beta and the partial correlation depends on the correlations among the predictors, and between predictors and the dependent variable:

$$r^2_{XA \cdot B \ldots n} = \beta^2_{XA \cdot B \ldots n} \left(\frac{1 - R^2_{A \cdot B \ldots n}}{1 - R^2_{X \cdot B \ldots n}}\right) \qquad \text{or}$$

$$\beta^2_{XA \cdot B \ldots n} = r^2_{XA \cdot B \ldots n} \frac{(1 - R^2_{X \cdot B \ldots n})}{(1 - R^2_{A \cdot B \ldots n})}$$

X = dependent variable
$A, B \ldots n$ = independent variables
r = partial correlation
R = multiple correlation

Beta will exceed the partial correlation whenever the predictor in question is itself more predictable from the other predictors than is the dependent variable.

6.5   Some Other Multiple Regression Techniques

It has been noted that the MCA program can be viewed as a multiple regression technique using dummy variables. There exist a variety of computer programs for the computation of multiple regression. Since the basic results (the adjusted coefficients) are mathematically identical with those of the MCA program, after simple algebraic transformation, the relative merits of one program or another have to do with (a) convenience of input and output, and (b) the provision of certain ancillary statistics.

The MCA program requires no conversion of the basic data, no creation on card or input tape of "dummy variables." Each class of each predicting characteristic becomes in essence a dummy variable. Most regression programs would require a separate recoding to create the variables. On the other hand, some regression programs will take both numerical and categorical predictors, and in some situations one may want to create a rather special set of predictor variables incorporating interaction effects anyway.

Some regression programs will introduce the explanatory variables one at a time, recalculating the regression each time, so the user can tell what the introduction of a new variable does to all the estimated coefficients. Some programs even determine by a programmed set of rules the order in which variables are to be introduced, and even try dropping variables. These stepwise procedures are most appropriate with numerical variables, or when they can treat each *set* of categories together, but it makes little sense to introduce single dummy variables one at a time and recalculate the whole regression, particularly when one adds an age class, then an education class, and then an occupational class.

6.6   Sampling Errors

The major differences in what is computed for any one regression or MCA computation are that most regression programs calculate sampling errors of each regression coefficient and the MCA program does not. The statistical logic behind these estimated errors does not fit most actual samples since typical samples are not simple random but clustered and multi-stage. On the other hand, work by Kish and others at the Survey Research Center shows that if one estimates these sampling errors by balanced replicates which compare half-samples properly taken in relation to the sample design, they are closely related to the ones computed by regression programs. There is a small "sample design effect" which makes them larger than simple random sampling errors by an average of about six percent.[5]

There is a simpler way of looking at the individual coefficients: they can be

----

[5]The method is described briefly in Kish (1965, page 586) and in Kish (1957, page 164). It is developed more fully in McCarthey (1966), Kish and Frankel (1970) and Frankel (1971).

viewed not as regression coefficients with a sampling error, but as each representing a mean value for a subgroup of the sample, with some adjustments made to it. For simple random samples, the sampling error of a mean is inversely proportional to the number of cases on which it is based, and directly proportional to some basic variability, usually represented by the standard deviation. Since the subgroups vary greatly in size, in most cases most of the variability in the sampling errors among coefficients in one analysis may come from the subgroup size, rather than from small differences in subgroup variability. As one test of this, we estimated the sampling errors of 39 coefficients using 12 half-sample replications (eliminating coefficients based on fewer than 100 cases). The simple correlation between these properly estimated sampling errors and a crude approximation using the overall standard deviation divided by the square root of the number of cases with that characteristic was:

$$\text{Sampling error of adjusted deviation (coefficient)} = -.04 + 1.05 \, (\sigma_y / \sqrt{N})$$

$$R^2 = .73$$

Interestingly enough, the approximation is much worse for the unadjusted subgroup means, though the same relationship holds:

$$\text{Sampling error of unadjusted deviation (subgroup mean)} = .01 + 1.02 \, (\sigma_y / \sqrt{N})$$

$$R^2 = .47$$

Kish and Frankel have done some experiments with replication estimates of the sampling stability of the etas, betas, and the unadjusted and adjusted coefficients. They prefer to fit the sampling errors to the form:

$$\text{Standard error} = \frac{a}{\sqrt{N}}$$

allowing the data to show how close a comes to the standard deviation. The fits are quite good, with the a for unadjusted subcell means (or deviations) within a few percentage points of the overall standard deviation, and the a for the adjusted coefficients within a few percentage points of the standard deviation times the square root of $1-R^2$, i.e., the standard deviation of the residuals from the whole regression (Kish and Frankel 1966).

One test case does not prove that such a simple approximation will always be appropriate. In particular, it will depend on whether the predictor classes vary greatly in size (so that it is the N rather than the variability that matters most), and on whether they cut across the sample design in such a way as to lead to large and varied "design effects" on their stability.

Similar comparisons with similar results have been made estimating the simple random sampling errors not by regression equation estimates, but by simple random sample halves (Kish and Frankel 1966).

Where the coefficients are expressed as deviations from an excluded class, as in regression with dummy variables, and the mean of the excluded class is unknown and may be extremely large or small, the relation of the coefficients to their standard errors is misleading anyway. This can be solved by translating the coefficients into the form in which the MCA program gives them. But in any case, what most researchers are interested in is the importance or significance of each *set* of subclasses as a group, that is the importance of age or education, or of one particular age or education group. And since, with large samples, almost anything that is important is also significant (though not always vice versa), measures of the importance of each predictor taken as a whole set of coefficients are what matter most. Only the MCA program computes such measures. These are the etas for each set of unadjusted deviations and the betas for each set of adjusted deviations, thinking of them as a new scale.

Estimates of the sampling errors of the etas and betas have been made by the method of replication for the analysis just mentioned, and they turned out to have coefficients of variation around 0.2 (Kish and Frankel 1966).

6.7   The Advantages of MCA Printout

Finally, there is the matter of the form in which the results are presented and printed out. We feel that the basic form—all coefficients expressed as deviations from the mean, not from some unknown mean of the excluded class in each set—is far better for presentation and analysis. The constant term in the predicting equation is the overall mean, not some composite sum of means of the excluded subclasses. And the adjusted and unadjusted subgroup means are available in the same table. Since the differences between the unadjusted and adjusted deviations (or means) are an indication of the amount of intercorrelation the analysis adjusted for, it is useful to be able to compare them easily. And when adjustments are extensive, it is useful to look for the correlations between predictors that account for them. In this regard, the presentation of the intercorrelations among the predictors, not as two-way relations between dummy variables, but as two-way tables of the sets of subclasses, is far easier to use.

And when the intercorrelations are so high as to be causing real instability in the estimates, this can be traced by following the iterations to see which coefficients are still being adjusted in the later iterations. The calculated standard errors of the individual dummy variable regression coefficients are no substitute for this, although they should be large in such cases.

All in all, then, MCA's major advantage is in taking the data the way they usually come, and printing out the results one is most likely to want to present in a convenient way.

Appendices

Appendix A

Set-Up Instructions

MCA is part of the Institute for Social Research's OSIRIS package of computer programs.[1] MCA control card formats and data formats conform to OSIRIS conventions. These are described in *OSIRIS III Volume 1: System and Program Design* (The University of Michigan 1973).

Input
1. Dictionary, on cards, tape, or disk (in OSIRIS format).
2. Data, on cards, tape, or disk (in OSIRIS format).
3. Program control cards.

Output
1. Printout. The contents of the printout are listed in Appendix B.
2. A dataset of residuals written on tape or disk (optional for each analsis). The dataset is written in standard OSIRIS format and consists of the output dictionary and the data file. The data file contains, for each case, an ID variable, a dependent variable, a calculated value, a residual value and, if one was specified, a weight variable. The characteristics of a residuals dataset are as follows:

Variable No.	Name	Field Width	No. of Dec.	MD1 Code
1	Interview number	Same as input	Same as input	Same as input
2	Observed value	Same as input	Same as input	Same as input
3	Calculated value	Same as input observed value	Same as input observed value	9's to fill field

---

[1]OSIRIS is a set of programs and supporting subroutines. The OSIRIS computer software system was jointly developed by the component Centers of the Institute for Social Research, The University of Michigan, using funds from the NSF, the Inter-university Consortium for Political Research and other sources.

Variable No.	Name	Field Width	No. of Dec.	MD1 Code
4	Observed-calculated value	Same as input observed value	Same as input observed value	9's to fill field
5	Weight (if weight was specified)	Same as input	Same as input	Same as input

If the observed value is missing data, or the case was excluded by filtering, outlier, or maximum code checking, a residual record is output with variables 3 and 4 set to 9's.

Restrictions

1. The total number of variables must be less than or equal to 200. This total includes local filter variables and and V variables which are used in OSIRIS recode statements[2] but not in MCA.

2. There must be one and only one dependent variable per analysis packet; it must be measured on an interval scale, or be dichotomous, and should not be badly skewed. It may have up to 7 digits, with any number of decimal places, and may have positive or negative values.

3. There may be 1-50 predictor variables per analysis packet; each must be categorized, preferably with 5-7 categories. The categories must have integer codes in the range 0-31, unless only one predictor is used, in which case it may have codes in the range 0-2999.

4. It is not possible to use the maximum number of predictors, each with the maximum number of categories in a given analysis packet. See the formula for the computer core requirement given in section 3.5 of this monograph.

Missing Data

Cases with missing data values for the dependent variable may be eliminated from an analysis using the local parameter MDOPTION. (In an OSIRIS dataset, missing data codes are stored in a machine readable "dictionary" which is part of the dataset.) Cases with missing data values for the predictors may be eliminated from a single analysis using a local filter card or from all analyses using a global filter card. (Using filters to exclude cases with missing data is useful *only* if the missing data codes are in the range 0-31; if, for a case, the value for any predictor is greater than 31, the case is automatically

---

[2]Recode statements are described in *OSIRIS III Volume 1: System and Program Design*.

excluded from all analyses in the run.)[3] Cases with missing data values for the weight variable will be eliminated automatically.

Executing the Program

The Job Control Language (JCL), OSIRIS monitor control cards, and program control cards needed to execute MCA are outlined below. Cards must be supplied in the indicated order. Details on the OSIRIS monitor and its catalogued procedure and on OSIRIS recode statements are given in *OSIRIS III Volume I: System and Program Design.* The "xxxx" and "yyyy" in the ddnames that follow are determined by the INFILE global parameter and the OUTFILE local parameter.

//	EXEC	OSIRIS
//DICTxxxx	DD	Define the input dictionary.
		(Omit this DD card if $DICT is used.)
//DATAxxxx	DD	Define the input data.
		(Omit this DD card if $DATA is used.)
//DICTyyyy	DD	Define the output residuals dictionary file.
		(Omit this DD card if residuals not requested.)
//DATAyyyy	DD	Define the output residuals data file.
		(Omit this DD card if residuals not requested.)
//DICTzzzz	DD	Define the second output residuals dictionary file.
		(Omit this DD card if a second residuals dataset is is not requested.)
//DATAzzzz	DD	Define the second output residuals data file.
		(Omit this DD card if a second residuals dataset is not requested.)
		(2 DD cards are needed for each residuals output.)
//SETUP	DD	*

$RUN MCA
$RECODE (Optional.)[4]
   Recode statements.
$SETUP
  1. Global filter. (Optional.)
  2. Label card.
  3. Global parameters.

---

[3]See section 2.1b. Often it is a good idea *not* to eliminate cases with missing data on predictors.

[4]A powerful recoding subroutine is one of the OSIRIS features available to MCA. For MCA users it is particularly useful for creating pattern variables (see the COMBINE keyword) and for recoding predictors so that they have codes in the range 0-31.

4.   Analysis packets. (As many as desired.)
    a. Local filter. (Optional.)
    b. Local label card.
    c. Local parameters.
    d. Predictor variable list(s).
$DICT (Optional.)
  Dictionary cards.
$DATA (Optional.)
  Data cards.
/*

Program Control Cards

1. Global filter. (Optional .) A filter card is used to select a subset of cases by specifying certain values of certain variables. A global filter applies to the whole program run. (In contrast, a local filter applies only to a single analysis packet.) Normally a filter is used to select a subset of cases, e.g. urban males. Note that it can also be used to screen out observations which are contained in a pair of overlapped predictor categories.

  Examples: (a) INCLUDE V2=1-5 AND V7=23,27,35 AND V8=1,3,4,6*
            (b) EXCLUDE V10=2-3,6,8-9 AND V30=001-004 OR
                V91=025*

The rules are:
—Punch INCLUDE or EXCLUDE, beginning anywhere on the first card. Continue punching on the first card; if it is necessary to use more than 1 card, end the first card with a comma or conjunction (AND or OR) and continue on the next card.
—Maximum number of expressions per run: 15. (An expression includes V, the variable number, an equals sign, and a list of values.)
—An asterisk must end the list of expressions.
—Variables may appear in any order and in more than one expression.
—Expressions may be connected by AND or OR:
  a. AND indicates that the values in all connected expressions must be found in order to select the case.
  b. OR indicates the case will be selected if any or all of the specified values are found.
  c. AND expressions are executed before OR expressions.
—Values specified must have the field width of the variable, i.e. lead zeros must be punched.
—Values may be specified singly, separated by commas, or in a range (e.g. 001-004).
—Values may be positive or negative, but a value may not vary from a negative value to a positive value. (If necessary, separate the values into two ranges.)

—Negative ranges should be expressed as in example (a) for global filter, and as in (b) for local filter:
  (a) V1 = -01 -- 10,000-009*
  (b) V1 = -10 -- 1,0-9*

2. Label card. One card containing up to 80 characters to label the printed output.

3. Global parameters. Parameters are chosen from those described below, must be separated by blanks, commas or both, and must be ended with an asterisk. Defaults are in italics. If all parameters are allowed to default, a card with just an asterisk must be supplied.

INFILE=*IN*/xxxx Allows the user to specify a 1-4 character input ddname suffix.

PRINT=*DICT*/NODICT DICT: Print the input dictionary.

BADDATA=*STOP*/SKIP/MD1/MD2 If non-numeric characters (including embedded blanks, &'s, and -'s and all-blank fields) are found in numeric variables, the program should:
STOP: Terminate the run.
SKIP: Skip the case.
MD1: Recode a full field of & to a full field of nines plus 1 (i.e., recode & to 10, && to 100, etc.). Recode a full field of - to a full field of nines plus 2 (i.e., recode - to 11, -- to 101, etc.) Recode all other non-numeric values to the first missing data code.
MD2: Recode full fields of & and - as specified in MD1 above. Recode all other non-numeric values to the second missing data code.
For SKIP, MD1, and MD2, a message is printed about the number of cases so treated.
MD1 and MD2 refer to "missing data codes." They are specified, for each variable, in the OSIRIS dictionary for the dataset. They are explained in *OSIRIS III Volume 1: System and Program Design.*

BADPACKETS=*STOP*/GO If some analysis packets contain errors:
STOP: End the run after all control cards are checked.
GO: Run analyses for packets which had no errors.
* End the global parameters with an asterisk.

4. Analysis packets. As many packets as desired may be supplied.

  a. Local filter. (Optional.) Selects a subset of cases for one analysis. The format for a filter is described under 1 above.

b. Local label card. One card containing up to 80 characters to title the individual analysis.

c. Local parameters. Parameters are chosen from those described below. Defaults are in italics. If all parameters are to default, a card with just an asterisk must be supplied.

PRINT=(*NORMAL*/TABLES/HISTORY/ALL)

        NORM: No special printing.

        TABL: Print the pairwise cross-tabulation of the predictors.

        HIST: Print the coefficients from all iterations.

        ALL: Print pairwise cross-tabulations of the predictors and print coefficients from all iterations.

WEIGHT=(variable number, maxcode) (Optional.) if this keyword does not occur, it is assumed the data are unweighted. If weighting, the variable number of the weight variable should be specified. Maxcode is the maximum code for the weight variable (the range is 0-9999999). Cases with weight values which exceed maxcode are excluded from the analysis. Default for the maxcode is 9999999.

DEPVAR=(variable number, maxcode) Variable number is the variable number of the dependent variable. Maxcode is the maximum code for the dependent variable (range is 0-9999999). The variable number must always be specified. Default for the maxcode is 9999999.

MDOPTION=*BOTH*/MD1/MD2/NONE This parameter governs the handling of cases that have missing data on the *dependent* variable. The program should:

        BOTH: Eliminate cases where the dependent variable equals MD1 or is in the range of MD2.

        MD1: Eliminate cases where the dependent variable equals MD1.

        MD2: Eliminate cases where the dependent variable in the range of MD2.

        NONE: Do not eliminate any cases on the basis of missing data.

        (MD1 and MD2 refer to missing data codes which are defined for each variable in the OSIRIS dictionary for the dataset; they are explained in *OSIRIS III Volume 1:System and Program Design*.)

OUTLIERS=*INCLUDE*/EXCLUDE
>INCL: Include in the analysis cases in which the value of the dependent variable is an outlier.
>EXCL: Exclude from the analysis cases in which the value of the dependent variable is an outlier.

OUTDISTANCE=*5*/n Number of standard deviations from the global mean used to define an outlier.

ITERATIONS=*25*/n The maximum number of iterations. Range = 1-99999.

TEST=%*MEAN*/CUTOFF/%RATIO/NONE The convergence test desired:
>%MEA: Test whether the change in all coefficients from one iteration to the next is below a specified fraction of the grand mean.
>CUTO: Test whether the change in all coefficients from one iteration to the next is less than a specified value.
>%RAT: Test whether the change in all coefficients from one iteration to the next is less than a specified fraction of the ratio of the standard deviation of the dependent variable to its mean.
>NONE: The program will iterate until the maximum number of iterations has been exceeded.

CRITERION=.*005*/n Supply a numeric value which is the tolerance of the convergence test selected. Range=0.0 to 1.0 (Punch decimal point.)

RUNS=*1*/n
>Set n equal to the number of analyses included in this packet, i.e., to the number of predictor variable lists which will be supplied.
>This option is not available if a one-way analysis of variance is being done (i.e., only 1 predictor is specified).

NOSUPPRS/SUPPRESS
>SUPP: Suppress printing of residuals.

OUTFILE=*OUT*/yyyy Allows the user to specify a 1-4 character residuals output ddname suffix.

NORESIDUAL/RESIDUAL/SRESIDUAL
>NORE: No residuals.
>RESI: Produce residuals: apply the MCA model only to cases passing global and local filter, missing-data, outlier, and maximum-code checking; pad other cases with 9's in the predicted value, residual and ID fields.

SRES: Produce residuals: apply the MCA model (developed for the subset of cases passing local filter, missing data, outlier and maximum-code checking) to all cases passing the global filter.[5]

Notes: Residuals cannot be obtained if only 1 predictor variable is specified, i.e., if a one-way analysis of variance is being done.
If more than one analysis is done within a packet, residuals may be computed only for the first.

IDVAR=variable number The number of an identification variable to be included in the residuals dataset. The default is the sequential number of the case within the set of cases for which residuals are calculated.

*   End the local parameters with an asterisk.

d. Predictor variable lists. For each analysis desired for this packet, supply a separate list of the variables to be used as predictors.
Examples: (a) V1-V6,V9,V16,V20-V120,V18,V11,V209*
          (b) V2,V5,V7,V10,V12,V15,V21,V26,V29,V31-V45,V52,
          V67-V92,V115,V136* (continuation card)

The rules are:
—Columns 1-80 may be used.
—Punching is free format (i.e., blanks may appear anywhere).
—An asterisk must follow the last item on the last card.
—Variables may be specified in any order.
—Single variables or variable ranges (e.g., V31-V45) must be separated by commas.
—V must precede *every* variable number (e.g., V31-45 is not valid).
—If the list must be continued, stop with a comma, and continue punching anywhere on the next card. As many continuation cards as needed may be used.

---

[5]Caution: In order to obtain residuals by applying the model developed on a subset of cases to a larger set of cases, the subset must contain all the predictor codes which appear in the larger set. For example, if the value 3 appears for predictor 1 in the larger set, then it is necessary that the value 3 appear for predictor 1 in the subset. Otherwise a coeffieient won't be developed for predictor 1, value 3. MCA does not check that this requirement is met; if violated, MCA produces erroneous, but possibly interpretable, residual values. If SRES is specified, the user should check the condition by obtaining marginals for the predictors, on both the large set of cases and the subset of cases, prior to the MCA run.

Appendix B

Reading the Printout
and Presenting Results

Several previous sections of this monograph are relevant to interpreting MCA printout: the user should see section 1.5 for short descriptions of the major statistics printed by the program, section 4.1 for the formulas used to compute these statistics, and section 6.3 for a discussion of the interpretation of betas. This Appendix covers additional material. In section B.1, items which appear on the printout are listed; separate lists are given for a MCA analysis and a one-way analysis of variance. In each, items are listed in the order in which they appear on the printout, with an indication of which are optional. Additionally, a note is included on how to read floating point notation. Section B.2 discusses the interpretation of adjusted coefficients, the crux of the additive model on which MCA is based. And, finally, in section B.3, some suggestions are given for presenting MCA results.

B.1 The Printout

The printout from the MCA program is different, depending on whether a MCA analysis or a one-way analysis of variance was done.

B.1a Printout for a MCA analysis (2 or more predictors). The first page of the printout documents the global settings for the entire run. Following this, the input dictionary is printed (unless the option PRINT=NODICT was specified). The number of cases passing the global filter is printed. (Only printed for the first analysis of the run.)

For each analysis requested, the following are printed:

—Parameters for the analysis, including the number of cases omitted for various reasons (e.g., because of missing data on the dependent variable).

—Frequency tables. (Optional.) These are weighted if a weight variable was specified.

—Convergence history. (Optional.) If multicollinearity is a problem, examination of the convergence history is a good way to pin-point the vari-

ables causing trouble. Instability of coefficients during the final iterations is a sign of multicollinearity. The coefficients for each predictor
are printed in a row.

—Statement that coefficients converged or that they did not converge.

—Documentation of the last and next to last coefficients. (Printed only if
the convergence test was not met.) The difference between the final two
coefficients is printed for each predictor. Predictors which are inter-
correlated, and thus causing trouble, will have relatively large differences.

—Dependent variable (Y) statistics.
  –grand mean
  –standard deviation
  –sum of dependent variable (Y)
  –sum of dependent variable (Y) squared
  –total sum of squares
  –explained sum of squares
  –residual sum of squares
  –number of cases used in the analysis
  –sum of the weights
Except for number of cases, these items are weighted if a weight variable
was specified. Otherwise each case is given a weight of 1.

—Predictor category statistics. These statistics are computed for each class
of the predictor; when the class is empty, the entire line of statistics is
omitted. They are:
  –class code
  –number of cases in class
  –weighted percentage of sample used that falls into class
  –class mean
  –deviation of class mean from grand mean
  –adjusted deviation, or coefficient (solution to normal equations)
  –adjusted mean (grand mean plus adjusted deviation)
  –standard deviation of dependent variable for class
The last five items are weighted if a weight variable was specified. See
section B.2 below for a discussion of adjusted coefficients.

—Predictor summary statistics. Six summary statistics are computed for
each predictor, representing its adjusted and unadjusted contribution
to explaining the variance of the dependent variable. They are:
  –eta
  –eta squared
  –beta
  –beta squared

–unadjusted sum of squares, i.e., unadjusted contribution to explaining the variance of the dependent variable

–adjusted sum of squares, i.e., adjusted contribution to explaining the variance of the dependent variable

These statistics are all weighted if a weight variable was specified. See section 6.3 for a discussion of the interpretation of beta and beta squared. Eta squared is sometimes called the correlation ratio.

—Analysis summary statistics. After the last predictor, three additional statistics are printed. The adjustment factor for the degrees of freedom is also printed. The summary printout consists of:

–multiple correlation coefficient, squared, unadjusted

–adjustment factor

–multiple correlation coefficient, adjusted

–multiple correlation coefficient, adjusted, squared

These, except for the adjustment factor, are weighted if a weight variable was specified.

—Residuals (Optional.) If a residuals printout is requested it will consist of the residuals dictionary and, for each case, the following:

–interview number

–observed value

–calculated value

–residual

–weight, if a weight variable was specified

—Residual summary statistics. (If residuals are requested, summary statistics are printed regardless of whether or not the residuals themselves are printed.)

–sum of weights of residuals

–mean of residuals

–variance of residuals

–skewness of residuals

–kurtosis of residuals

B.1b   Printout for a one-way analysis of variance (1 predictor). The first page of the printout documents the global settings for the entire run. Following this, the input dictionary is printed (unless the option PRINT=NODICT was specified). The number of cases passing the global filter is printed. (Only printed for the first analysis of the run.)

For each analysis requested, the following are printed:

—Parameters for the analysis, including the number of cases eliminated for various reasons (e.g., because of missing data on the dependent variable).

—Predictor category statistics. These statistics are computed for each class of the predictor; when the class is empty, the entire line of statistics is omitted. They are:
  –class code
  –number of cases in class
  –weighted percentage of sample used that falls into class
  –class mean
  –standard deviation of dependent variable for the class
  –sum of dependent variable (Y) for class
  –sum of dependent variable (Y) expressed as a percent
  –sum of dependent variable (Y) squared for class
The last six items are weighted if a weight variable was specified.

—Totals. Totals over all predictor categories are printed for each of the above predictor category statistics (except class code).

—One way analysis of variance summary statistics.
  –unadjusted eta^2
  –adjustment factor[1]
  –adjusted eta
  –adjusted eta^2
  –total sum of squares
  –between means sum of squares
  –within groups sum of squares
  –F value (degrees of freedom are printed)
If the data are weighted these items, except for the last one, are weighted, and the F value is not printed. See the first paragraph of section 4.2 and the footnote it references for a discussion of the problem of interpreting an F statistic when using weighted data. If an F test is desired it can be computed from the basic statistics printed by the program.

B.1c Reading floating point notation. MCA statistics are printed using a Fortran "G" format. This means that normally numbers will be printed in their conventional form (e.g. 36.0554); however, if a number is very large or very small, the program will automatically switch to "floating point" notation. A number written in floating point can be recognized by an "E" exponent to the right of the number. The correct decimal place is indicated after the E by a two-digit number indicating the number of places that the decimal

---

[1] The formula which the program uses to adjust eta (for capitalization on chance in fitting the model) in a one-way analysis is the same as the formula used to adjust the multiple correlation coefficient in a Multiple Classification Analysis. (See section 3.4 for a discussion of the adjustment procedure and section 4.1 for the formula.) We have not made a rigorous examination of the legitimacy of applying the formula to eta. We have, however, done a number of empirical tests which indicate that the adjustment is about right when the true relationship is known to be close to zero.

point should be moved. If the indicator is preceded by a minus sign, the decimal belongs a corresponding number of places to the left;if there is no sign, the decimal place should be moved to the right. For example:

3.60554 E-01 = .360554

3.60554 E 02 = 360.554

B.2 Adjusted Coefficients

The major interpretation in a MCA is of the adjusted and unadjusted co-efficients printed out for each subclass. In a population where there was no correlation among the predictors, the observations in *one* class of characteristic A would be distributed over all classes of the other characteristics in a fashion identical to the way in which those in other classes of A were distributed. Hence, the unadjusted mean Y for each subclass of A would be an unbiased estimate of the effect of belonging to that class of characteristic A. In the real world, however, characteristics are correlated. Young people are more likely to be in lower income groups, and in higher education groups than are older people. The multivariate process is essentially one of adjusting for these "non-orthogonalities." The adjusted means are estimates of what the mean would have been if the group had been exactly like the total population in its distribution over all the other predictor classifications. It is useful not only to have the "pure" effects of each class adjusted for all the other characteristics, but also to see how these adjusted effects differ from the unadjusted effects.

Both the adjusted and unadjusted coefficients are expressed by the program as deviations from the overall mean, and are constrained so that their sum, weighted by the proportion in each subclass, is zero.

The adjusted coefficients for any predictor may be considered an estimate of the effect of that predictor alone "holding constant" all other predictors in the analysis. Differences between the adjusted and unadjusted coefficients can be analyzed, and explanations for these differences may often be found in the two-way tables of predictors. It is often valuable to compare the coefficients within a predictor to see whether there is a pattern or, possibly, a lack of pattern which is of theoretical interest.

The coefficients for the predictors do not provide definitive information about logical priorities, chains of causation, or about interaction effects. It is possible for the program to assign considerable explanatory power to a variable late in a causal chain, such as an attitude, when much of the credit "really" belongs to a logically prior, but not as powerful variable, such as race.[2]

Interaction effects of two or more predictors on the dependent variable will

---

[2]For a careful examination of the relationship between statistics and causal analysis, see Blalock (1964).

not be revealed by the program, since the assumption is that the effects of all the predictors are additive, i.e. the effect for predictor A is assumed to be the same for one class of predictor B as it is for every other class.

A difficulty in using the adjusted coefficients as a presentational device is that the additivity assumptions may lead to absurd adjusted means for some groups (less than zero, for instance) if the assumption is inappropriate for the data being analyzed. This is particularly likely when the dependent variable is a dichotomy, such as home ownership. Clearly, it is not sensible to predict that less than 0 percent of a subgroup own a home.

B.3 Presentation of Results

It is most informative to the reader to present first the etas and betas, measures of the relative importance of each predictor singly and in competition with the others, and then to present the unadjusted and adjusted subgroup averages, together with a detailed description of what the subclasses represent and with the number of cases in each. (The number of cases should be included because it is an indicator of the potential variability of the estimates.) Multiple R^2 unadjusted and multiple R^2 adjusted are also usually reported.

We recommend that the results be given in the form of unadjusted and adjusted subgroup averages rather than in the form of deviations because the user finds it easier to scan unadjusted and adjusted subgroup averages than positive and negative deviations. However, the adjusted deviations can be included for convenience in seeing the net effects of each predictor. As noted above, a complication of subgroup averages is that occasionally the expected value is impossible (e.g. negative although the dependent variable is a variable with no negative values); if impossible expected values are presented, a short explanatory note should be included.

A word about significance tests. The pure theory of statistical inference assumes that one keeps careful track of degrees of freedom. Every calculation and test uses up degrees of freedom, and hence successive tests of alternative regression models raise real problems. Since it is common, tempting, and even sensible to try a succession of analyses on the same set of data, most of us end up unable to apply the usual significance tests appropriately. Yet it would be absurd to restrict the analysis to the testing of a single regression model, even with the added flexibility that dummy variables allow.

The most usual suggestion, rarely followed, is to reserve an independent sample on which the finally selected model can be tested, both for significance, and for the stability of the parameters compared with the other ransacked sample. If that is not done, the results should be presented as the best set of hypotheses one was able to derive from the set of data, ready to be tested on some other set of data.

Examples of presentation of MCA results can be found in Barfield and Morgan (1969), Blumenthal, Kahn, Andrews and Head (1972), Johnston and Bachman (1972), Johnston (1973), Katona, Strumpel and Zahn (1971), Morgan, David, Cohen and Brazer (1962), Mueller (1969), and Pelz and Andrews (1966).

Appendix C

MCA Macro Flow Chart

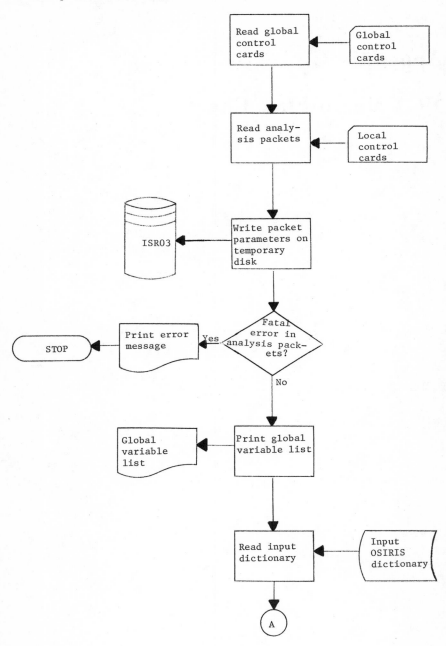

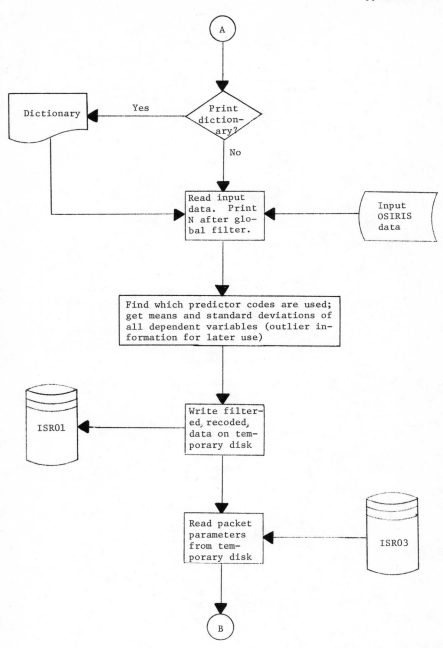

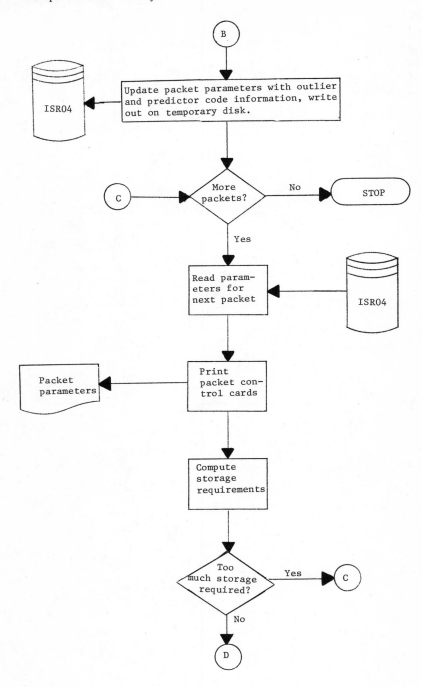

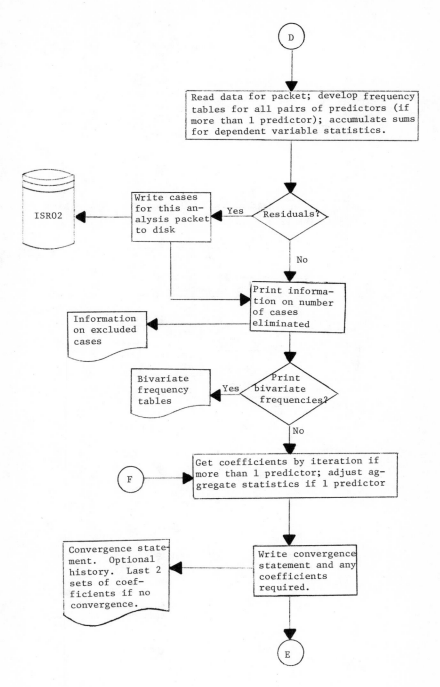

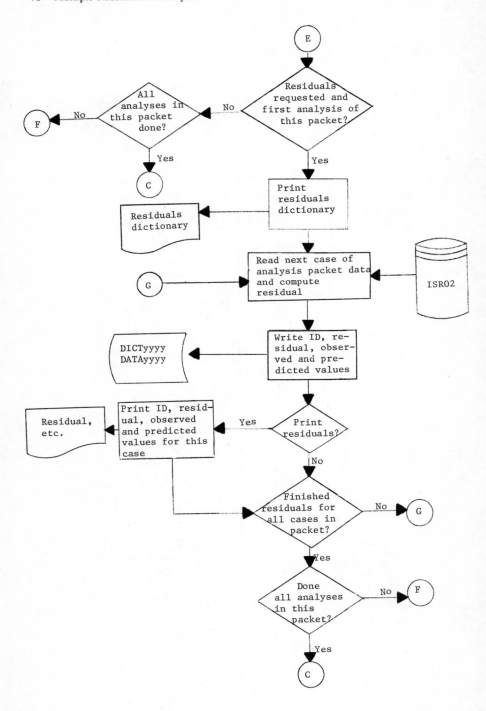

Appendix D

Sample Output

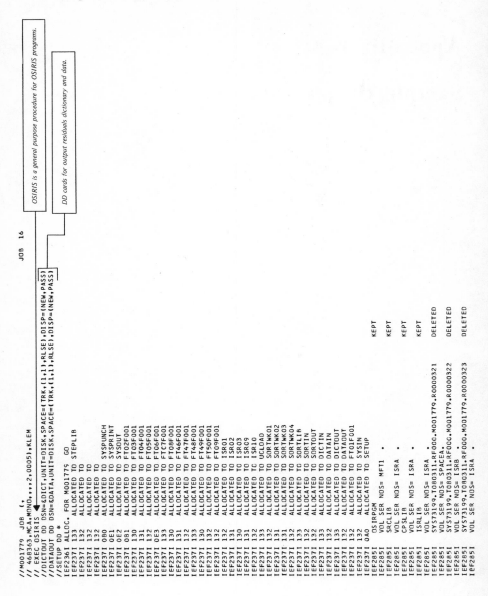

JOB 16

OSIRIS is a general purpose procedure for OSIRIS programs.

DD cards for output residuals dictionary and data.

```
//M001779  JOB  (,
// 468363,MCA,MONO,,,,2,00051,KLEM
// EXEC OSIRIS
//DICTOUT DD DSN=&DICT,UNIT=DISK,SPACE=(TRK,(1,1),RLSE),DISP=(NEW,PASS)
//DATAOUT DD DSN=&DATA,UNIT=DISK,SPACE=(TRK,(1,1),RLSE),DISP=(NEW,PASS)
//SETUP DD *
IEF236I ALLOC. FOR M001779 GO
IEF237I 133  ALLOCATED TO STEPLIB
IEF237I 132  ALLOCATED TO
IEF237I 132  ALLOCATED TO
IEF237I 132  ALLOCATED TO
IEF237I 080  ALLOCATED TO SYSPUNCH
IEF237I 0E1  ALLOCATED TO SYSPRINT
IEF237I 0E2  ALLOCATED TO SYSOUT
IEF237I 081  ALLOCATED TO FT02F001
IEF237I 130  ALLOCATED TO FT03F001
IEF237I 131  ALLOCATED TO FT04F001
IEF237I 132  ALLOCATED TO FT05F001
IEF237I 0E3  ALLOCATED TO FT06F001
IEF237I 133  ALLOCATED TO FTC7F001
IEF237I 130  ALLOCATED TO FT08F001
IEF237I 131  ALLOCATED TO FT46F001
IEF237I 132  ALLOCATED TO FT47F001
IEF237I 133  ALLOCATED TO FT48F001
IEF237I 130  ALLOCATED TO FT49F001
IEF237I 132  ALLOCATED TO FT50F001
IEF237I 132  ALLOCATED TO FT09F001
IEF237I 132  ALLOCATED TO ISR01
IEF237I 131  ALLOCATED TO ISR02
IEF237I 130  ALLOCATED TO ISR03
IEF237I 131  ALLOCATED TO ISR09
IEF237I 132  ALLOCATED TO ISR10
IEF237I 133  ALLOCATED TO UCLOAD
IEF237I 132  ALLOCATED TO SORTWK01
IEF237I 131  ALLOCATED TO SORTWK02
IEF237I 131  ALLOCATED TO SORTWK03
IEF237I 132  ALLOCATED TO SORTWK04
IEF237I 133  ALLOCATED TO SORTLIB
IEF237I 132  ALLOCATED TO SORTIN
IEF237I 132  ALLOCATED TO SORTOUT
IEF237I 133  ALLOCATED TO DICTIN
IEF237I 132  ALLOCATED TO DATAIN
IEF237I 133  ALLOCATED TO DICTOUT
IEF237I 132  ALLOCATED TO DATAOUT
IEF237I 132  ALLOCATED TO FT01F001
IEF237I 132  ALLOCATED TO SYSIN
IEF237I 0A0  ALLOCATED TO SETUP
IEF285I       OSIRPGM                                    KEPT
IEF285I       VOL SER NOS= MFT1  .
IEF285I       SRCLIB                                     KEPT
IEF285I       VOL SER NOS= ISRA  .
IEF285I       CPSLIB                                     KEPT
IEF285I       VOL SER NOS= ISRA  .
IEF285I       ISRLIB                                     KEPT
IEF285I       VOL SER NOS= ISRA  .
IEF285I       SYS73199.T080311.RF000.M001779.R0000321   DELETED
IEF285I       VOL SER NOS= SPACEA,
IEF285I       SYS73199.T080311.RF000.M001779.R0000322   DELETED
IEF285I       VOL SER NOS= ISRB  .
IEF285I       SYS73199.T080311.RF000.M001779.R0000323   DELETED
IEF285I       VOL SER NOS= ISRA  .
```

```
IEF285I    SYS73199.T080311.RF000.M001779.R0000324    DELETED
IEF285I    VOL SER NOS=
IEF285I    SYS73199.T080311.RF000.M001779.R0000325    DELETED
IEF285I    VOL SER NOS= MFT1 .
IEF285I    SYS73199.T080311.RF000.M001779.R0000326    DELETED
IEF285I    VOL SER NOS= SPACEA.
IEF285I    SYS73199.T080311.RF000.M001779.R0000327    DELETED
IEF285I    VOL SER NOS= ISRB .
IEF285I    SYS73199.T080311.RF000.M001779.R0000328    DELETED
IEF285I    VOL SER NOS= ISRA .
IEF285I    SYS73199.T080311.RF000.M001779.R0000329    DELETED
IEF285I    VOL SER NOS= MFT1 .
IEF285I    SYS73199.T080311.RF000.M001779.R0000330    DELETED
IEF285I    VOL SER NOS= SPACEA.
IEF285I    ISRNEWS                                     KEPT
IEF285I    VOL SER NOS= ISRA .
IEF283I    SYS73199.T080311.RF000.M001779.R0000328    NOT DELETED 8
IEF285I    VOL SER NOS= ISRA 1.
IEF283I    SYS73199.T080311.RF000.M001779.R0000331    DELETED
IEF285I    VOL SER NOS= ISRA .
IEF285I    SYS73199.T080311.RF000.M001779.R0000332    DELETED
IEF285I    VOL SER NOS= ISRB .
IEF285I    SYS73199.T080311.RF000.M001779.R0000333    DELETED
IEF285I    VOL SER NOS= SPACEA.
IEF285I    SYS73199.T080311.RF000.M001779.R0000334    DELETED
IEF285I    VOL SER NOS= ISRB .
IEF285I    SYS73199.T080311.RF000.M001779.R0000335    DELETED
IEF285I    VOL SER NOS= ISRA .
IEF283I    SYS73199.T080311.RF000.M001779.R0000325    NOT DELETED 8
IEF285I    VOL SER NOS= MFT1 1.
IEF283I    SYS73199.T080311.RF000.M001779.R0000331    NOT DELETED 8
IEF285I    VOL SER NOS= ISRA 1.
IEF283I    SYS73199.T080311.RF000.M001779.R0000332    NOT DELETED 8
IEF285I    VOL SER NOS= ISRB 1.
IEF283I    SYS73199.T080311.RF000.M001779.R0000334    NOT DELETED 8
IEF285I    VOL SER NOS= ISRB 1.
IEF283I    SYS73199.T080311.RF000.M001779.R0000335    NOT DELETED 8
IEF285I    VOL SER NOS= ISRA 1.
IEF285I    SYS1.SORTLIB                                KEPT
IEF285I    VOL SER NOS= MFT1 .
IEF285I    SYS73199.T080311.RF00C.M001779.R0000328    PASSED
IEF285I    VOL SER NOS= ISRA .
IEF285I    SYS73199.T080311.RF000.M001779.R0000328    PASSED
IEF285I    VOL SER NOS= MFT1 .
IEF285I    SYS73199.T080311.RF000.M001779.R0000329    PASSED
IEF285I    VOL SER NOS= MFT1 .
IEF285I    SYS73199.T080311.RF000.M001779.R0000328    PASSED
IEF285I    VOL SER NOS= ISRA .
IEF285I    SYS73199.T08C311.RF00C.M001779.DICT         PASSED
IEF285I    VOL SER NOS= MFT1 .
IEF285I    SYS73199.T080311.RF00C.M001779.DATA         PASSED
IEF285I    VOL SER NOS= ISRA .
IEF285I    SYS73199.TC83311.RF000.M001779.R0000336     DELETED
IEF285I    VOL SER NOS= ISRA .
IEF283I    SYS73199.T080311.RF00C.M001779.R0000336     NOT DELETED 8
IEF285I    VOL SER NOS= ISRA 1.
IEF285I    SYS73199.T080311.RF000.M001779.R0000337     DELETED
IEF285I    VOL SER NOS=
ISR011I    STEP GO       EXECUTION TIME =  066.72 SEC.
IER013I    PARTITION    2: SIZE= 104, LWM=FOF000, HWM=FOF000 , CORE ALLOCATED= 100, CORE USED=  90
IEF285I    SYS73199.T08C311.RF000.M001779.R0000328     KEPT
```

```
IEF285I    VOL SER NOS= ISRA .
IEF285I    SYS73199.TC8C311.RF000.M001779.R0000328
IEF285I    VOL SER NOS= ISRA .
IEF285I    SYS73199.T080311.RF00C.M001779.R0000329
IEF285I    VOL SER NOS= MFT1 .
IEF285I    SYS73199.T080311.RF000.M001779.R0000328
IEF285I    VOL SER NOS= ISRA .
IEF285I    SYS73199.T080311.RF000.M001779.DICT
IEF285I    VOL SER NOS= MFT1 .
IEF285I    SYS73199.T080311.RF000.M001779.DATA
IEF285I    VOL SER NOS= ISRA .
ISR012I TOT. M001779  EXECUTION TIME =   066.72 SEC.
ISR016I TIME OF DAY = 10.00.25, DATE = 73.199

                                            KEPT

                                            KEPT

                                            KEPT

                                            DELETED

                                            DELETED
```

```
*****          INSTITUTE FOR SOCIAL RESEARCH MONITOR SYSTEM          02/05/73          *****

******TIME IS   9:55: 9

******LISTING OF SET-UP FOLLOWS:

         CARD            1         2         3         4         5         6         7         8
         NO.    1234567890123456789012345678901234567890123456789012345678901234567890123456789 0
          1     $RUN MCA
          2     2 MCA ANALYSES, DATA ARE FROM PELZ AND ANDREWS(1966).
          3     *
          4     ANALYSIS 1:   VARIABLE 2 IS DEPENDENT.
          5     DEPV=(2,99)   OUTLIER=EXCL OUTDISTANCE=3*
          6     V3,V6*
          7     INCLUDE V4=0.0-3.0*
          8     ANALYSIS 2. NOTE USE OF LOCAL FILTER. RESIDUALS REQUESTED. VAR. 1 IS DEPENDENT.
          9     DEPV=(1,99) RESI*
         10     V3,V6*
         11     $DICT
         12     $PRINT
         13        3    1   6    1
         14     T   1 TECH CONTRIB        1 1       2
         15     T   2 USEFULNESS          1         3
         16     T   3 VARIABLE 3          3         4
         17     T   4 ADJ B REPORTS       7         8
         18     T   5 VARIABLE 5          8        91
         19     T   6 VARIABLE 6         10        11
         20     $DATA                    11
         21     $PRINT
         22     25222843822
         23     76833823524
         24     61593842922
         25     09063842922
         26     85733833724
         27     31311843123
         28     15364842922
         29     93894834624
         30     99944825824
         31     66713823824
         32     02611834022
         33     02012842772
         34     52434843322
         35     95962834424
         36     89804831824
         37     9C793833823
         38     16172842922
         39     22193843322
         40     72853833824
         41     13112842322
         42     23983211824
         43     47965834124
         44     62523823924
         45     37272843822
         46     66453842323
         47     50693833924

         CARD            1         2         3         4         5         6         7         8
         NO.    1234567890123456789012345678901234567890123456789012345678901234567890123456789 0
```

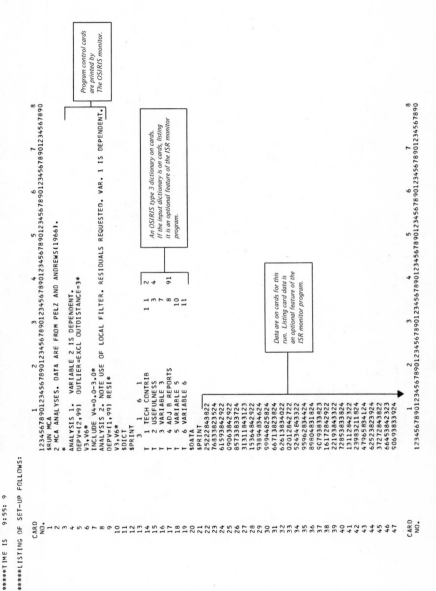

Program control cards
are printed by
The OSIRIS monitor.

An OSIRIS type 3 dictionary is on cards.
If the input dictionary is on cards, listing
it is an optional feature of the ISR monitor
program.

Data are on cards for this
run. Listing card data is
an optional feature of the
ISR monitor program.

```
                   1         2         3         4         5         6         7         8
         1234567890123456789012345678901234567890123456789012345678901234567890123456789 0
CARD
NO.
 48      31254843823
 49      99892834324
 50      93834836024
 51      83734836324
 52      19362842323
 53      87772834424
 54      79924823824
 55      76572833424
 56      02012840722
 57      96804833524
 58      72863842924
 59      69562843823
 60      48995021924
 61      76814014324
 62      54584061922
 63      38145043523
 64      84904013524
 65      40565021924
 66      16193043922
 67      46374021922
 68      30333041922
 69      92934043523
 70      79845043524
 71      71494043023
 72      24164033023
 73      31243041923
 74      17133044322
 75      44914023524
 76      48224044123
 77      06633041923
 78      06054043922
 79      99995044624
 80      21584034324
 81      30483041922
 82      68593045124
 83      81756033023
 84      45524035024
 85      38253043923
 86      51324043023
 87      54444041923
 88      41773023524
 89      97975044723
 90      63375043923
 91      15052041922
 92      47293044623
 93      54912820724
 94      72935021924
 95      26393041923
 96      49492842323
 97      34364843123

CARD
NO.
                   1         2         3         4         5         6         7         8
         1234567890123456789012345678901234567890123456789012345678901234567890123456789 0
```

```
                                  1         2         3         4         5         6         7         8
                         1234567890123456789012345678901234567890123456789012345678901234567890123456789012345678901234567890
CARD
NO.
 98    11374820724
 99    41654823824
100    67924833124
101    83775823824
102    36514021922
103    99954031924
104    88964031924
105    79975023524
106    11254011924
107    23163031923
108    13304011924
109    19075041923
110    25515021924
111    20302630823
112    81464633023
113    54613630824
114    74893620824
115    29215640824
116    24344642423
117    87923636624
118    39192643223
119    48283624824
120    17252623424
121    28543640822
122    35124644523
123    69495544722
124    79694634224
125    94793622424
126    16141633623
127    35573613024
128    19753632824
129    52654643623
130    42264663923
131    79694624724
132    35494642823
133    56733633024
134    69964641923
135    89844644024
136    99733643922
137    46264644223
138    37633623224
139    81943633924
140    57483643423
141    83965523024
142    23332623223
143    98484643323
144    72752632824
145    16563622442
146    09824632424
147    59655630824

CARD                              1         2         3         4         5         6         7         8
NO.                      1234567890123456789012345678901234567890123456789012345678901234567890123456789012345678901234567890
```

```
                    1         2         3         4         5         6         7         8
           12345678901234567890123456789012345678901234567890123456789012345678901234567890
CARD       40052641922
NO.        35924620824
148        18754623024
149        92233633924
150        59494614722
151
152
```

MCA -- OSIRIS MULTIPLE CLASSIFICATION ANALYSIS PROGRAM -- JULY 6, 1973

2 MCA ANALYSES. DATA ARE FROM PELZ AND ANDREWS(1966).

*

DEPV=(2,99) OUTLIER=EXCL OUTDISTANCE=3*

THE VARIABLE LIST IS:

V3,V6*

THE FOLLOWING IS THE LOCAL FILTER FOR THIS ANALYSIS

INCLUDE V4=0.0-3.0*

DEPV=(1,99) RESI*

THE VARIABLE LIST IS:

V3,V6*

THE GLOBAL VARIABLE LIST IS:

3   6   2   1   4

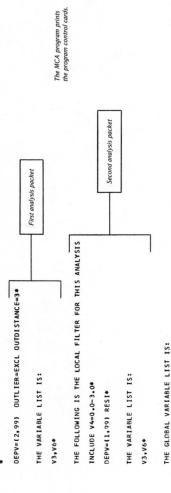

First analysis packet

Second analysis packet

*The MCA program prints
the program control cards.*

VAR.	TYPE	VARIABLE NAME	TLOC	WIDTH	NODEC	RESP.	MDCODE1	MDCODE2	REFNO	ID	TSEQNO
T 1	0	TECH CONTRIB	1	2	0	1					
T 2	0	USEFULNESS	3	2	0	1					
T 3	0	VARIABLE 3	7	1	0	1					
T 4	0	ADJ B REPORTS	8	2	1	1					
T 6	0	VARIABLE 6	11	1	0	1					

TOTAL N (AFTER APPLICATION OF THE TOTAL FILTER,IF REQUESTED) IS 131

MCA, optionally, prints the OSIRIS dictionary.

ANALYSIS PACKET: ANALYSIS 1.   VARIABLE 2 IS DEPENDENT.

*****TIME 9 57 33 33

DEPENDENT VARIABLE

   NAME   USEFULNESS
   SUBSCRIPT   2   *
   MAX.CODE   99.000000
   INCLUDE MD1?   NO
   INCLUDE MD2?   NO
   INCLUDE OUTLYERS? NO
   RANGE : L.T.   -31.014679
          G.T.   142.22076

WEIGHT VARIABLE?   NO

PRINT FREQUENCIES?   NO

ITERATION MAXIMUM   25

CONVERGENCE TEST   2

TEST FOR CONVERGENCE   0.00500

PRINT COEFFICIENTS?   NU

NUMBER OF PREDICTORS   2

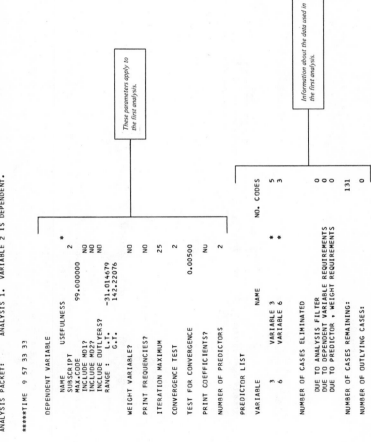

These parameters apply to the first analysis.

PREDICTOR LIST

VARIABLE   NAME   NO. CODES

  3   VARIABLE 3   *   5
  6   VARIABLE 6   *   3

NUMBER OF CASES ELIMINATED

  DUE TO ANALYSIS FILTER   0
  DUE TO DEPENDENT VARIABLE REQUIREMENTS   0
  DUE TO PREDICTOR , WEIGHT REQUIREMENTS   0

NUMBER OF CASES REMAINING: 131

NUMBER OF OUTLYING CASES: 0

Information about the data used in the first analysis.

RESULTS BASED ON TEST   2   ITERATION   5

Successful convergence.

DEPENDENT VARIABLE STATISTICS

DEPENDENT VARIABLE (Y)   =   2: USEFULNESS

M E A N   =   55.603043

STANDARD DEVIATION   =   28.872574

SUM OF Y   =   7284.0000

SUM OF Y SQUARE   =   513384.00

TOTAL SUM OF SQUARES   =   108371.38

EXPLAINED SUM OF SQUARE   =   46457.914

RESIDUAL SUM OF SQUARES   =   61873.461

NUMBER OF CASES   =   131

PREDICTOR CATEGORY STATISTICS

PREDICTOR   3:   VARIABLE   3

CLASS	NO OF CASES	SUM OF WEIGHTS	PER CENTS	CLASS MEAN	UNADJUSTED DEVIATION FROM GRAND MEAN	COEFFICIENT	ADJUSTED MEAN	STAND DEV.
1	7	7	5.3	61.4286	5.8255157	-4.8369179	50.766113	28.941649
2	29	29	22.1	69.6207	14.017639	2.8545961	58.457626	22.975785
3	37	37	28.2	65.4595	13.856415	4.9301653	60.533203	23.718950
4	57	57	43.5	39.2807	-16.322342	-3.7489719	51.854065	26.460722
6	1	1	0.8	26.0000	-29.603043	-17.645920	37.957123	0.0

ETA-SQUARE = 0.26854044   BETA-SQUARE = 0.22256110E-01
      ETA = 0.5182C886   BETA = 0.14918482

UNADJUSTED DEVIATION SS = 29102.098
ADJUSTED DEVIATION SS = 2411.9255

Eta-square indicates that approximately twenty-seven percent of the variance in the dependent variable is explainable by variable 3.

Beta-square is an approximate measure of the relationship between variable 3 and the dependent variable (Usefulness) while holding constant variable 6.

PREDICTOR   6:   VARIABLE   6

CLASS	NO OF CASES	SUM OF WEIGHTS	PER CENTS	CLASS MEAN	UNADJUSTED DEVIATION FROM GRAND MEAN	COEFFICIENT	ADJUSTED MEAN	STAND DEV.
2	27	27	20.6	31.7778	-23.825272	-21.091171	34.511871	22.030282
3	39	39	29.8	41.5641	-14.038940	-11.883295	43.719742	24.228946
4	65	65	49.6	73.9231	18.320023	15.890882	71.493912	21.145558

ETA-SQUARE = 0.41365576   BETA-SQUARE = 0.31310576
      ETA = 0.64316076   BETA = 0.55955857

UNADJUSTED DEVIATION SS = 44828.449
ADJUSTED DEVIATION SS = 33931.707

Approximately forty-one percent of the variance in the dependent variable is explainable by variable 6.

The beta-square for variable 6 is larger than the beta-square for variable 3 indicating that variable 6 holding constant variable 3 is more important than variable 3 holding constant variable 6.

ANALYSIS SUMMARY STATISTICS

R-SQUARED (UNADJUSTED) = PROPORTION OF VARIATION EXPLAINED BY FITTED MODEL = 0.42906

ADJUSTMENT FOR DEGREES OF FREEDOM = 1.04839

***MULTIPLE R (ADJUSTED) = 0.63359   MULTIPLE R-SQUARED (ADJUSTED) = 0.40144

Estimate of how much variance the obtained coefficients would explain if used in an additive model applied to a different (but comparable) set of cases—e. g. the population from which the sample analyzed was drawn.

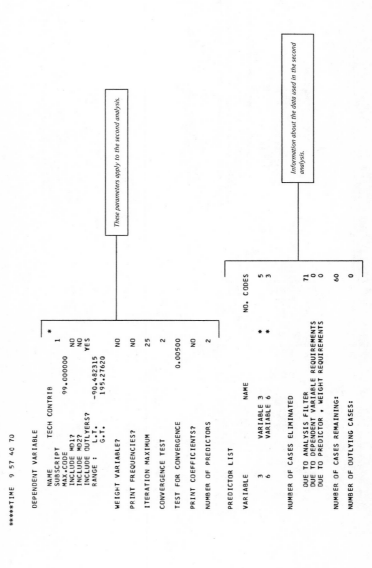

```
ANALYSIS PACKET:      ANALYSIS 2. NOTE USE OF LOCAl FILTER. RESIDUALS REQUESTED. VAR. 1 IS DEPENDENT.

*****TIME  9  57  40  70

DEPENDENT VARIABLE

   NAME      TECH CONTRIB
   SUBSCRIPT           1
   MAX.CODE     99.000000
   INCLUDE MD1?        NO
   INCLUDE MD2?        NO
   INCLUDE OUTLYERS?  YES
   RANGE :  L.T.  -90.482315
            G.T.  195.27620

WEIGHT VARIABLE?             NO
PRINT FREQUENCIES?           NO
ITERATION MAXIMUM            25
CONVERGENCE TEST              2
TEST FOR CONVERGENCE    0.00500
PRINT COEFFICIENTS?          NO
NUMBER OF PREDICTORS          2
```

These parameters apply to the second analysis.

```
PREDICTOR LIST

VARIABLE       NAME          NO. CODES

   3       VARIABLE 3           5   *
   6       VARIABLE 6           3   *

NUMBER OF CASES ELIMINATED

   DUE TO ANALYSIS FILTER                       71
   DUE TO DEPENDENT VARIABLE REQUIREMENTS        0
   DUE TO PREDICTOR , WEIGHT REQUIREMENTS        0

NUMBER OF CASES REMAINING:                      60

NUMBER OF OUTLYING CASES:                        0
```

Information about the data used in the second analysis.

RESULTS BASED ON TEST 2 ITERATION 4

Successful convergence.

DEPENDENT VARIABLE STATISTICS

DEPENDENT VARIABLE (Y) = 1: TECH CONTRIB

MEAN = 41.399994

STANDARD DEVIATION = 27.242920

SUM OF Y = 2484.0000

SUM OF Y SQUARE = 146626.00

TOTAL SUM OF SQUARES = 43788.438

EXPLAINED SUM OF SQUARE = 11651.125

RESIDUAL SUM OF SQUARES = 32137.313

NUMBER OF CASES = 60

PREDICTOR CATEGORY STATISTICS

PREDICTOR   3: VARIABLE   3

CLASS	NO OF CASES	SUM OF WEIGHTS	PER CENTS	CLASS MEAN	UNADJUSTED DEVIATION FROM GRAND MEAN	COEFFICIENT	ADJUSTED MEAN	STAND DEV.
1	4	4	6.7	20.5000	-20.899994	-31.452652	9.9473419	11.000000
2	14	14	23.3	46.5714	5.1714325	0.55824280E-01	41.455811	26.022813
3	14	14	23.3	58.1429	16.742859	10.698002	52.097992	32.820859
4	28	28	46.7	33.4286	-7.9714355	-0.88367748	40.516312	21.767166

ETA-SQUARE = 0.17870915   BETA-SQUARE = 0.12745923
ETA = 0.42274004   BETA = 0.35701436

UNADJUSTED DEVIATION SS =   7825.3945
ADJUSTED DEVIATION SS =   5581.2422

PREDICTOR   6: VARIABLE   6

CLASS	NO OF CASES	SUM OF WEIGHTS	PER CENTS	CLASS MEAN	UNADJUSTED DEVIATION FROM GRAND MEAN	COEFFICIENT	ADJUSTED MEAN	STAND DEV.
2	16	16	26.7	25.8125	-15.587494	-14.879972	26.520020	17.874446
3	18	18	30.0	41.6111	0.21110535	-2.1223497	39.277634	23.926699
4	26	26	43.3	50.8461	9.4461517	10.626224	52.026215	30.367999

ETA-SQUARE = 0.14177924   BETA-SQUARE = 0.14980042
ETA = 0.37653583   BETA = 0.38704062

UNADJUSTED DEVIATION SS =   6208.2930
ADJUSTED DEVIATION SS =   6555.5273

ANALYSIS SUMMARY STATISTICS

R-SQUARED (UNADJUSTED) = PROPORTION OF VARIATION EXPLAINED BY FITTED MODEL = 0.26608

ADJUSTMENT FOR DEGREES OF FREEDOM =   1.09259

***MULTIPLE R (ADJUSTED) = 0.44511   MULTIPLE R-SQUARED (ADJUSTED) = 0.19812

FOLLOWING IS A LISTING OF THE RESIDUALS DICTIONARY

	VAR.	TYPE	VARIABLE NAME	TLOC	WIDTH	NODEC	RESP.	MDCODE1	MDCODE2	REFNO	ID	TSEQNO
T	1	0	IDENTIFYING VAR.	1	7	0	1	9999999	9999999			
T	2	0	OBSERVED VALUE	8	2	0	1	99	99			
T	3	0	PREDICTED VALUE	10	2	0	1	99	99			
T	4	0	RESIDUAL	12	3	0	1	999	999			

IDENTIFYING VARIABLE	OBSERVED VALUE	PREDICTED VALUE	RESIDUAL	WEIGHT VARIABLE
1.000	9999999.000	9999999.000	9999999.000	
2.000	9999999.000	9999999.000	9999999.000	
3.000	61.000	25.636	35.364	
4.000	9.000	25.636	-16.636	
5.000	9999999.000	9999999.000	9999999.000	
6.000	15.000	25.636	-10.636	
7.000	9999999.000	9999999.000	9999999.000	
8.000	9999999.000	9999999.000	9999999.000	
9.000	9999999.000	9999999.000	9999999.000	
10.000	9999999.000	9999999.000	9999999.000	
11.000	2.000	25.636	-23.636	
12.000	9999999.000	9999999.000	9999999.000	
13.000	9999999.000	9999999.000	9999999.000	
14.000	89.000	62.724	26.276	
15.000	9999999.000	9999999.000	9999999.000	
16.000	16.000	25.636	-9.636	
17.000	9999999.000	9999999.000	9999999.000	
18.000	9999999.000	9999999.000	9999999.000	
19.000	13.000	25.636	-12.636	
20.000	23.000	20.574	2.426	
21.000	9999999.000	9999999.000	9999999.000	
22.000	9999999.000	9999999.000	9999999.000	
23.000	9999999.000	9999999.000	9999999.000	
24.000	66.000	38.394	27.606	
25.000	9999999.000	9999999.000	9999999.000	
26.000	9999999.000	9999999.000	9999999.000	
27.000	9999999.000	9999999.000	9999999.000	
28.000	9999999.000	9999999.000	9999999.000	
29.000	9999999.000	9999999.000	9999999.000	
30.000	19.000	38.394	-19.394	
31.000	9999999.000	9999999.000	9999999.000	
32.000	9999999.000	9999999.000	9999999.000	
33.000	9999999.000	9999999.000	9999999.000	
34.000	2.000	25.636	-23.636	
35.000	9999999.000	9999999.000	9999999.000	
36.000	72.000	51.143	20.857	
37.000	9999999.000	9999999.000	9999999.000	
38.000	48.000	52.082	-4.082	
39.000	54.000	25.636	28.364	
40.000	9999999.000	9999999.000	9999999.000	
41.000	9999999.000	9999999.000	9999999.000	
42.000	40.000	52.082	-12.082	
43.000	9999999.000	9999999.000	9999999.000	
44.000	46.000	26.576	19.424	
45.000	30.000	25.636	4.364	
46.000	9999999.000	9999999.000	9999999.000	
47.000	9999999.000	9999999.000	9999999.000	
48.000	71.000	38.394	32.606	
49.000	24.000	49.976	-25.976	
50.000	31.000	38.394	-7.394	
51.000	9999999.000	9999999.000	9999999.000	
52.000	9999999.000	9999999.000	9999999.000	
53.000	6.000	38.394	-32.394	
54.000	9999999.000	9999999.000	9999999.000	
55.000	9999999.000	9999999.000	9999999.000	
56.000	9999999.000	9999999.000	9999999.000	
57.000	9999999.000	9999999.000	9999999.000	
58.000	9999999.000	9999999.000	9999999.000	
59.000	9999999.000	9999999.000	9999999.000	

The printing of residuals can be suppressed if desired.

Seventy-one cases which did not pass the analysis filter have been padded with 9's.

60.000	30.000	25.636	4.364
61.000	9999999.000	9999999.000	9999999.000
62.000	81.000	49.976	31.024
63.000	9999999.000	9999999.000	9999999.000
64.000	9999999.000	9999999.000	9999999.000
65.000	51.000	38.394	12.606
66.000	54.000	38.394	15.606
67.000	9999999.000	9999999.000	9999999.000
68.000	9999999.000	9999999.000	9999999.000
69.000	9999999.000	9999999.000	9999999.000
70.000	15.000	25.636	-10.636
71.000	9999999.000	9999999.000	9999999.000
72.000	54.000	52.082	1.918
73.000	9999999.000	9999999.000	9999999.000
74.000	26.000	38.394	19.918
75.000	49.000	38.394	-12.394
76.000	9999999.000	9999999.000	9999999.000
77.000	11.000	52.082	10.606
78.000	9999999.000	9999999.000	9999999.000
79.000	9999999.000	9999999.000	9999999.000
80.000	36.000	26.576	-41.082
81.000	99.000	62.724	9.424
82.000	88.000	62.724	36.276
83.000	9999999.000	9999999.000	9999999.000
84.000	11.000	20.574	25.276
85.000	23.000	49.976	-9.574
86.000	13.000	20.574	-26.976
87.000	19.000	38.394	-7.574
88.000	25.000	52.082	-19.394
89.000	20.000	49.976	-27.082
90.000	81.000	62.724	-29.976
91.000	54.000	52.082	31.024
92.000	74.000	51.143	-8.724
93.000	29.000	38.394	21.918
94.000	24.000	25.636	-22.143
95.000	9999999.000	9999999.000	-14.394
96.000	9999999.000	9999999.000	9999999.000
97.000	9999999.000	9999999.000	9999999.000
98.000	9999999.000	9999999.000	9999999.000
99.000	28.000	25.636	2.364
100.000	9999999.000	9999999.000	9999999.000
101.000	9999999.000	9999999.000	9999999.000
102.000	9999999.000	9999999.000	9999999.000
103.000	94.000	52.082	41.918
104.000	9999999.000	9999999.000	9999999.000
105.000	35.000	20.574	14.426
106.000	19.000	62.724	-43.724
107.000	9999999.000	9999999.000	9999999.000
108.000	9999999.000	9999999.000	9999999.000
109.000	35.000	38.394	-3.394
110.000	9999999.000	9999999.000	9999999.000
111.000	96.000	62.724	33.276
112.000	9999999.000	9999999.000	9999999.000
113.000	69.000	38.394	30.606
114.000	9999999.000	9999999.000	9999999.000
115.000	9999999.000	9999999.000	9999999.000
116.000	9999999.000	9999999.000	9999999.000
117.000	9999999.000	9999999.000	9999999.000
118.000	9999999.000	9999999.000	9999999.000
119.000	9999999.000	9999999.000	9999999.000
120.000	83.000	52.082	30.918

121.000	9999999.000	9999999.000	9999999.000
122.000	9999999.000	9999999.000	9999999.000
123.000	72.000	62.724	9.276
124.000	16.000	26.576	-10.576
125.000	9.000	62.724	-53.724
126.000	59.000	62.724	-3.724
127.000	40.000	25.636	14.364
128.000	35.000	52.082	-17.082
129.000	18.000	52.082	-34.082
130.000	9999999.000	9999999.000	9999999.000
131.000	9999999.000	9999999.000	9999999.000

SUMMARY STATISTICS ON RESIDUALS

NO. CASES = 60.00000000

SUM OF WTS= 60.00000000

MEAN   =0.16784659924E-04

VARIANCE = 544.6323242

SKEWNESS =-.83204984466E-01

KURTOSIS =-.7661666870

******TIME 9 58 16 95 NORMAL TERMINATION OF MCA

Appendix E

Obtaining the Program and Adapting it to Other Systems

Obtaining Program

Information on obtaining MCA may be obtained from the Institute for Social Research program librarian by writing to:

Computer Program Librarian
Institute for Social Research
The University of Michigan
Ann Arbor, Michigan 48106

MCA is available either as part of the Institute for Social Research's statistical package for the IBM/360 (OSIRIS) or in a special AID/MCA package. The complete OSIRIS package consists of more than 40 programs and should be considered if the user plans extensive statistical calculations on his data. The AID/MCA package consists of only those two programs, the OSIRIS monitor, and the recode program, all in executable form.

Adaption

MCA is written to run on an IBM 360/40 under HASP/MFT with a Fortran G level compiler, IBM basic Assembler and 104K available byte partition. If the adaptation is to another IBM machine with the same language software, at least 104K core, and sufficient peripheral devices (a tape drive and 2314 disk unit or equivalent), no problems should be encountered if the instructions given with the package are followed carefully. Listed below are common adaptation problems and solutions. If the user's adaptation does not fall into one or more of these categories, it is suggested that a systems programmer be consulted.

Less than 104K

The simplest change here is to reduce the dimensionality of the T array (12000) appropriately. For example, if one had only 80K core available (24K too small) then a 6K full word reduction to the dimensionality of T is

99

sufficient, leading to T (6000) and a corresponding reduction in the number of predictors and predictor codes which can be used in an analysis packet. For the relevant formula see section 3.5—The Core Storage Limitation. If the dimensionality of the T array is changed, a test in the MCASB3 subroutine must also be changed.

Fortran E or Earlier Compiler

The basic problems here are non-recognition of the Fortran G level INTEGER * 2 specification and the "logical if" statement. Provided there is sufficient memory, all program INTEGER * 2 statements may be converted to INTEGER * 4. In addition, in OSIRIS subroutines, the GETDIC argument LIST must be changed to INTEGER * 4 and GETDIC reprogrammed. All "logical ifs" can be easily converted to their "arithmetic if" counterparts.

Inadequate Temporary Disk Storage Space

DD names ISR01 and ISR02 can be overridden in the JCL or changed in the program to specify tape units. Changing ISR01, which is used to store data passing the global filter, will substantially increase the I/O time as data will be read from tape for each analysis packet. Changing ISR02, which is used only if residuals are requested, will have the effects of saving disk space and increasing I/O time only if residuals are requested. (See Appendix C: MCA Macro Flow Chart.)

Non-IBM Machine

The Assembler coded subroutines GPIN and RCHAIO cannot be assembled on a non-IBM machine. Either these subroutines must be rewritten in the available Assembler language or the I/O and data manipulation portions of MCA must be reprogrammed.

Bibliography

Anderson, R. L. and Bancroft, T. A. *Statistical Theory in Research*. New York: McGraw Hill, 1952.

Andrews, F. M. and Messenger, R. C. *Multivariate Nominal Scale Analysis*. Ann Arbor, Michigan: Institute for Social Research, 1973.

Barfield, R. and Morgan, J. N. *Early Retirement: The Decision and The Experience*. Ann Arbor, Michigan: Institute for Social Research, 1969.

Bean, L. H. A simplified method of graphic curvilinear correlation. *Journal of the American Statistical Association*, 1929, *24*, 386-397.

Blalock, H. M. *Causal Inference in Non-Experimental Research*. Durham, North Carolina: University of North Carolina Press, 1964.

Blumenthal, M. D., Kahn, R. L., Andrews, F. M., and Head, K. B. *Justifying Violence: Attitudes of American Men*. Ann Arbor, Michigan: Institute for Social Research, 1972.

Darlington, R. B. Multiple regression in psychological research and practice. *Psychological Bulletin*, 1968, *69*, 161-182.

Ezekiel, M. and Fox, C. A. *Methods of Correlation and Regression Analysis*. 3rd ed. New York: Wiley, 1959.

Frankel, M. R. *Inference from Survey Samples: An Empirical Investigation*. Ann Arbor, Michigan: Institute for Social Research, 1971.

Goldberger, A. S. and Jochem, D. B. Note on stepwise least squares. *Journal of the American Statistical Association*, 1961, *56*, 105-110.

Hess, I. and Pillai, R. K. Multiple Classification Analysis. Ann Arbor, Michigan: Sampling Section, Survey Research Center, 1960. 16 pp.

Horst, P. Pattern analysis and configurational scoring. *Journal of Clinical Psychology*, 1954, *10*, 3-11.

Johnston, J. and Bachman, J. G. *Youth in Transition, Volume V: Young Men and Military Service*. Ann Arbor, Michigan: Institute for Social Research, 1972.

Johnston, L. *Drugs and American Youth*. Ann Arbor, Michigan: Institute for Social Research, 1973.

Kalton, G. A technique for choosing the number of alternative response categories to provide in order to locate an individual's position on a continuum. Three memos dated, Nov. 7, 1966, Feb. 10, 1967 and Mar. 10, 1967. Ann Arbor, Michigan: Sampling Section, Survey Research Center.

Katona, G., Strumpel B., And Zahn E. *Aspiration and Affluence: Comparative Studies in the United States and Western Europe*. New York: McGraw Hill, 1971.

Kempthorne, O. *The Design and Analysis of Experiments*. New York: Wiley, 1952.

Kish, L. Confidence intervals in clustered samples. *American Sociological Review*, 1957, *22*, 154-164.

Kish, L. *Survey Sampling*. New York: Wiley, 1965.

Kish, L. and Frankel, M. Standard errors of MCA multivariate regression coefficients and deviations. Mimeo. Ann Arbor, Michigan: Sampling Section, Survey Research Center, May 13, 1966.

Kish L. and Frankel, M. R. Balanced repeated replications for standard errors. *Journal of the American Statistical Association*, 1970, *65*, 1071-1094.

Lansing, J. B. and Morgan, J. N. *Economic Survey Methods*. Ann Arbor, Michigan: Institute for Social Research, 1971.

Lippitt, V. *Determinants of Consumer Demand for House Furnishings and Equipment*. Cambridge, Massachusetts: Harvard University Press, 1959.

Lubin, A. and Osborne, H. G. A theory of pattern analysis for the prediction of a quantitative criterion. *Psychometrika*, 1957, *22*, 63-73.

McCarthy, P. J. Replication: an approach to the analysis of data from complex surveys. Washington: National Center for Health Statistics, Series 2, No. 14, 1966.

Melichar, E. Least squares analysis of economic survey data. *Proceedings of the Business and Economic Statistics Section of ASA*, 1965.

Morgan, J. N. Consumer investment expenditures. *American Economic Review*, 1958, *48*, 874-902. (See especially the Appendix, pp. 898-901.)

Morgan, J. N., David, M., Cohen, W., and Brazer, H. *Income and Welfare in the United States*. New York: McGraw Hill, 1962.

Morgan, J. N. and Sonquist, J. Problems in the analysis of survey data and a proposal. *Journal of the American Statistical Association*, 1963, *58*, 415-435.

Mueller, E. *Technological Advance in An Expanding Economy: Its Impact on a Cross-Section of the Labor Force*. Ann Arbor, Michigan: Institute for Social Research, 1969.

Nunnally, J. C. *Psychometric Theory*. New York: McGraw Hill, 1967.

Pelz, D. C. and Andrews, F. M. The SRC computer program for multivariate analysis: some uses and limitations. Ann Arbor, Michigan: Survey Research Center, 1961. 46 pp. Publication #1844.

Pelz, D. C. and Andrews, F. M. *Scientists in Organizations: Productive Climates for Research and Development*. New York: Wiley, 1966.

Snedecor, G. W. *Statistical Methods*. Ames, Iowa: Iowa State College Press, 1946.

Sonquist, J. A. *Multivariate Model Building*. Ann Arbor, Michigan: Institute for Social Research, 1970.

Sonquist, J. A., Baker, E. L., and Morgan, J. N. *Searching for Structure*. (Rev. ed.) Ann Arbor, Michigan: Institute for Social Research, 1973. (This monograph supersedes Sonquist, J. A., and Morgan, J. N. *The Detection of Interaction Effects*. Ann Arbor, Michigan: Survey Research Center, 1964.)

Stevens, S. S. On the theory of scales of measurement. *Science*, 1946, *103*, 677-680.

Suits, D. B. Use of dummy variables in regression equations. *Journal of the American Statistical Association*, 1957, *52*, 548-551.

Sweeney, R. E. and Ulveling, E. F. A transformation for simplifying the interpretation of coefficients of binary variables in regression analysis. *The American Statistician*, 1972, *26*, 30-32.

The University of Michigan, Institute for Social Research. *OSIRIS III Volume I: System and Program Design*. Ann Arbor, Michigan, 1973.

Yates, F. The analysis of variance with unequal numbers in the different classes. *Journal of the American Statistical Association*, 1934, *29*, 51-66.

Index

Methodological Research Monographs
from the Institute for Social Research

OSIRIS: Architecture and Design. Judith Rattenbury and Neal Van Eck. 1973. 315 p.

Multivariate Nominal Scale Analysis: A Report on a New Analysis Technique and a Computer Program. Frank M. Andrews and Robert C. Messenger. 1973. 114 p.

THAID: A Sequential Analysis Program for the Analysis of Nominal Scale Dependent Variables. James N. Morgan and Robert C. Messenger. 1973. 98 p.

Multivariate Model Building: The Validation of a Search Strategy. John A. Sonquist. 1970. 264 p.

Searching for Structure. John A. Sonquist, Elizabeth Lauh Baker, and James N. Morgan. 1971. Revised edition, 1973. 248 p.

Multiple Classification Analysis: A Report on a Computer Program for Multiple Regression Using Categorical Predictors. Frank M. Andrews, James N. Morgan, John A. Sonquist, and Laura Klem. 1967. Revised edition, 1973. 105 p.

Inference from Survey Samples: An Empirical Investigation. Martin R. Frankel. 1971. 173 p.

SEPP: Sampling Error Program Package. Leslie Kish, Martin R. Frankel, and Neal Van Eck. 1972. 184 p.

Information about current prices and available editions may be obtained from the Sales Fulfillment Section, Institute for Social Research, The University of Michigan, Box 1248, Ann Arbor, Michigan 48106. Free catalogs and descriptive brochures are available on request.